25. JUN 11
07. JAN 12
25. FEB 12
HAMPSHIRE COUNTY LIBRARY
WITHDRAWN.

Hampshire
County Council

D1352376

C014707748

Trigonometry

Chapters 1 to 14 with answers to exercises

Christine Tootill

www.studymates.co.uk

ISBN: 978-1-84285-014-5

First published in 2010 by Studymates Limited.
PO Box 225, Abergele, LL18 9AY, United Kingdom.

Website: http://www.studymates.co.uk

Typeset by Vikatan Publishing Solutions, Chennai, India
Printed and bound in Europe

Studymates is a division of GLMP Ltd

Contents

Author's preface

Are you studying trigonometry? Then this book is certainly for you. By working through the topic from absolute basics to advanced concepts and methods, this book provides real help for students of mathematics aged 14 to adult who are working towards college examinations or following any other course at a similar level.

Trigonometry tends to cause more difficulties for students than many other topics in mathematics. This can be because students often have to move on to the next stage before having completely grasped the fundamental concepts. We have therefore included plenty of examples for you to use as practice in the early chapters as well as the later ones, and this of course is the best way to overcome these apparent difficulties.

This book takes you through a step-by-step approach and introduces you thoroughly to the topic of trigonometry. We have focused on the basics and taken as practical an approach as is possible in pure mathematics, so this book will enable you to understand the fundamental concepts and achieve success with examination questions.

Advice to students

Mathematics is a conceptual subject. It takes time to absorb new concepts, so don't try to rush through this book. It would be a very good idea to work through each chapter twice before you move on. Sometimes you may find you get stuck on a particular question, or you may not understand a particular point on first reading, but this doesn't mean you have to stop at that point.

Work through the rest of the chapter and then go back to whatever caused you a problem. If you still find it difficult, leave it, and move on. It is not always necessary to understand every word of a book in order to gain an overall understanding of the topic in hand. Also, it is not necessary to have answered every single example correctly before progressing to the next chapter. You will probably find that if you return to points of difficulty at a later date, you will by then have absorbed the ideas better and be able to cope with what previously seemed difficult. You are the only person who knows whether you have understood a topic and are ready to move on—so trust your instincts on this.

Do work with other students if you have the opportunity. It is an established fact that students can learn as much from each other as they can from teachers and books. Mutual support also dispels the notion, common to many students, that they are the only person who is finding the work hard and everyone else is cleverer than they are!

You will need a scientific calculator to do the work in this book, and from Chapter 8 onwards you will find a graphical calculator very useful.

Christine Tootill

1 First things first

One-minute overview

This chapter is a gentle introduction to the measurement of angles. We work first with degrees and then with radians. Even if you are not a beginner at trigonometry and you are confident about both of these methods of angle measurement, it would be a good idea to read through this chapter just to make yourself familiar with the style of this book.

Circular measurement

Degrees

Much of the everyday mathematics which we use today was developed in ancient times. The Babylonians, whose civilisation flourished around 2000 BC in the area which is modern-day Iraq, divided the day into 24 hours and each hour into 60 minutes. They also divided the circle into 360 degrees, and this has given us the foundation of our angle measurement system. They may have chosen the number 360 because it is close to the number of days in a year. Whether or not this is true, it was a good choice, because the number 360 has many factors, making it an easy number to work with.

So this is our starting point:

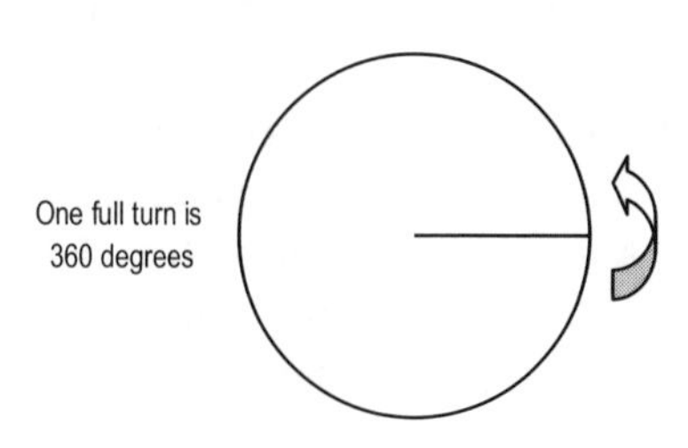

Fig 1.1

We use this symbol: ° for degrees.

One full turn is 360°.

A half turn is therefore 180°, and a quarter turn is 90°.

Examples (1.1)

The movement made by the minute hand on a traditional clock, is one full turn in one hour.

So in half an hour the minute hand moves through 180°, and in 15 minutes it moves through 90°.

The movement made by the hour hand is one full turn in 12 hours.

Direction of turn

We often need to indicate the direction of a turn. So we use the clock for this purpose. The hands of a clock turn in the *clockwise* direction. The arrow in Fig 1.1 is pointing in the *anticlockwise* direction.

Examples (1.2)

If you are facing North and you turn 90° clockwise, you are now facing East.

If you are facing South and you turn 90° anticlockwise, you are now facing East.

If you are facing South you can turn 180° clockwise or anticlockwise, and you will then be facing North.

Practice Questions (1.1)

Note – answers to all questions in the text are given in the Answers section at the end of the book.

1. Through how many degrees does the minute hand of a clock turn in 5 minutes?
2. If you are facing West and you turn 90° anticlockwise, which direction are you facing?
3. A beetle walks round the circle in Fig 1.1. It takes 20 minutes to complete the full turn. What is its speed in degrees per minute?

Practical measurement

For measurement and drawing, the instrument we need is a protractor. A protractor like the one in Fig 1.2 enables you to measure or draw angles between 0° and 180°.

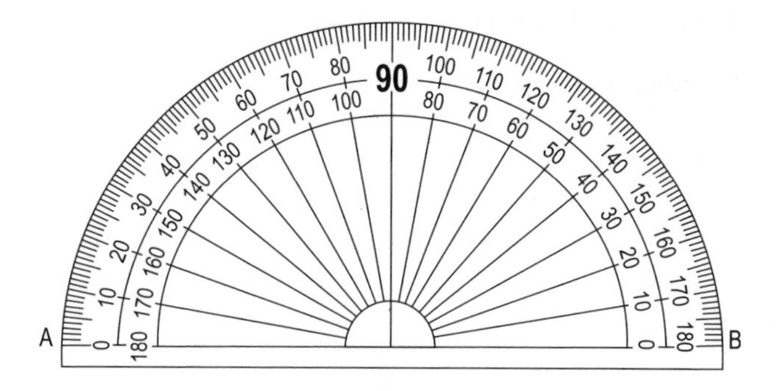

Fig 1.2

Examples (1.3)

To measure the angle between the two lines in Fig 1.3, we place the protractor so that the point at the centre of the line AB on the protractor is exactly at the point X where the two lines meet.

Then we turn the protractor so that the line AB on the protractor lies exactly on the line AX in Fig 1.2.

We find the size of the angle by reading the number on the outer edge of the protractor which the line XC is pointing to. We see that in this example the angle is 70°.

The notation for this angle is ∠AXC, or ∠CXA.

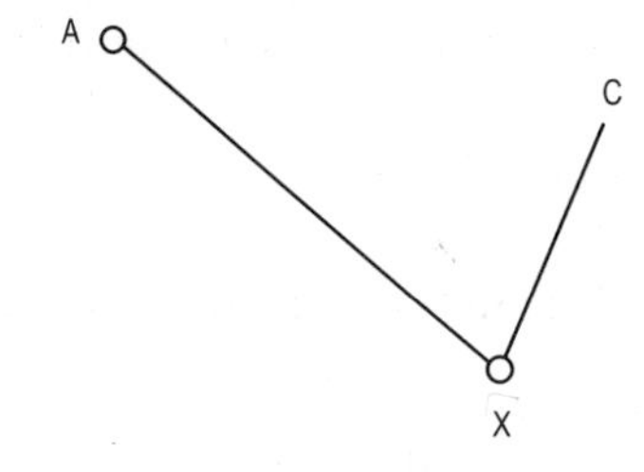

Fig 1.3

To construct a particular angle at either end of an existing straight line, lie the line AB of the protractor along the straight line with the point at the centre of AB exactly on the end of the straight line.

Then mark a point against the outer edge of the protractor at the required number of degrees.

Practice Questions (1.2)

1. Copy Fig 1.3 and extend the line CX to a point D so that DX is approximately the same length as CX.

   Measure $\angle AXD$. Write down the sum of $\angle AXD$ and $\angle AXC$.
2. Draw a line AB and construct an angle BAC = 48° and an angle ABD = 55°.

   Label the point E where the lines AC and AD meet. Measure $\angle AEB$.
3. Draw a line AB and construct a line BC so that $\angle ABC$ = 150°. Extend the line AB to a point D.

   Measure $\angle DBC$.

Some definitions

There are several words we use for angle sizes. You need to learn these terms if they are new to you:

An angle greater than 0° and less than 90° is an *acute* angle.

An angle of 90° is a *right* angle.

An angle greater than 90° but less than 180° is an *obtuse* angle.

An angle greater than 180° and less than 360° is a *reflex* angle.

Two angles which add up to 90° are called *complementary* angles.

Two angles which add up to 180° are called *supplementary* angles.

Practice Questions (1.3)

1. Complete this table by marking each angle A (acute), O (obtuse), or R (reflex):

65°	
165°	
265°	
184°	
101°	
86°	

2. Complete this table by marking each pair of angles C (complementary) or S (supplementary).
   If a pair of angles is neither of these, mark it N:

30°, 60°	
21°, 69°	
45°, 55°	
20°, 160°	
67°, 113°	
150°, 40°	

3. (a) How many degrees are there in:
   (i) 2 complete turns (ii) 3 complete turns (iii) 1½ turns
   (b) A turn of 380° is equivalent to a turn of 20°. Which angle is equivalent to a turn of 540°?

Fractions of angles

The Babylonians introduced the idea of dividing an angle into 60 "seconds".

Although one second is an extremely tiny angle, in precision work fractions of angles are important and must be calculated correctly.

The symbol for seconds is ". So an angle of 22½ degrees can be written as 22° 30".

Seconds were in common use until relatively recently, but nowadays it is more common to use decimals to indicate parts of angles. So for 22½ degrees we can write 22.5°.

In later work where you need to use your calculator, when your answer is not a whole number of degrees you should generally write your answer down giving one place of decimals.

Radians

Radians provide an alternative and quite different way of measuring angles.

One radian is the angle which is made by connecting the two ends of an arc to the centre of a circle, where the arc is the same length as the radius, as in Fig 1.4:

This diagram is not drawn to scale but it shows how an arc can be the same length as the radius:

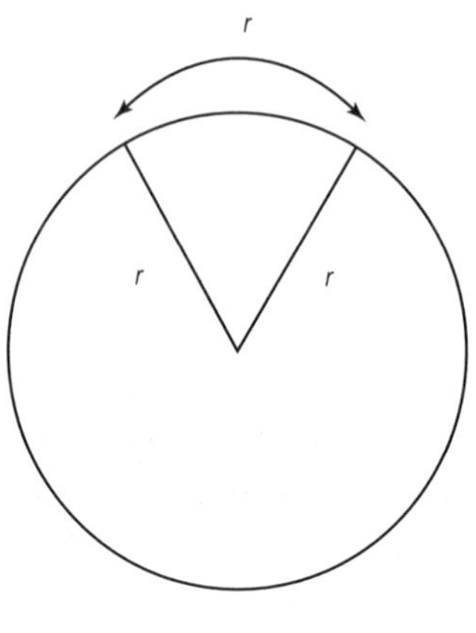

Fig 1.4

So the first question to ask is, how many radians are there in one full turn?

And the answer is not a whole number. This is because the length of the circumference of the circle is equal to

2π times the radius, r.

> If you are not familiar with, or have forgotten about, the circumference of a circle, you can revise this topic at: http://www.bbc.co.uk/schools/gcsebitesize/maths/shapes/circlesrev3.shtml

So the number of radians in one full turn =

$$\frac{\text{circumference}}{\text{radius}} = \frac{2\pi r}{r} = 2\pi$$

If this is new to you, learn this important fact now:

> There are 2π radians in one full turn.

The number π is approximately equal to 3.14, so 2π is approximately equal to 6.28.

Since there are more than 6 radians in a full turn, then one radian is less than 60°.

The symbol used for radians is: ^{C}

The symbol for "approximately equals" is: ≈

Practice Questions (1.4)

1. Using the mathematical symbols ^{C}, °, ≈, =, < and >, write these statements:
   (i) π is approximately equal to 3.14.
   (ii) One radian is less than 60 degrees.
   (iii) 360 degrees is more than 6 radians.
   (iv) 2π radians are equal to 360 degrees.
2. Using π, write down the number of radians equivalent to:
   (i) 180°
   (ii) 90°
   (iii) 2 full turns
   (iv) ¾ of one full turn

Tutorial

Progress Questions (1)

1. Angle α is an acute angle. Write down an expression for:
   (i) the angle which is complementary to angle α.
   (ii) the angle which is supplementary to angle α.
2. Look at your diagram for question 2 in Practice Questions (1.2).

   The points A, B, E form a triangle. $\angle BAE = 48°$, $\angle ABE = 55°$.

   $\angle AEB$ is the angle which you measured. Now write down these three angles and add them together.

   Your answer should be 180°. If your answer is 179° or 181°, then this small discrepancy is just due to slight inaccuracies in measuring or drawing. But if your answer is outside the range 179° to 181°, your drawing and measurement skills could be improved!

   But the point of this question is to remind you of this *very important fact:*

> The three angles of a triangle add up to 180°.

3. (i) Copy the triangle in Fig 1.5, with $\angle CAB = 90°$.
   It is usual to indicate a right angle with the mark ∟ in the appropriate position.

   (In this question, you can choose any sizes for the other angles)

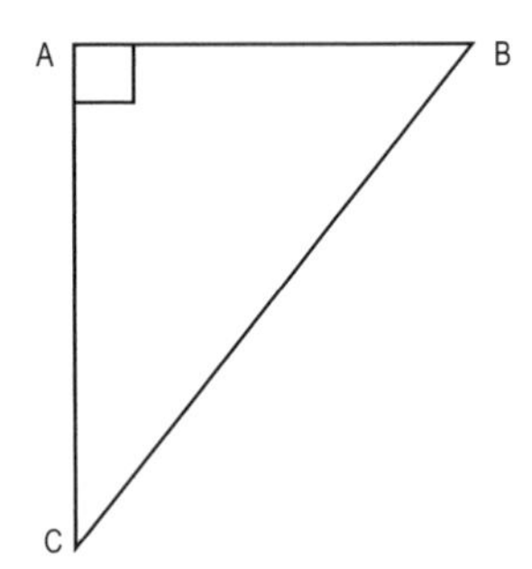

Fig 1.5

(ii) Mark the pair of complementary angles.

(iii) For these possible values of ∠ABC, write down the value of angle ∠ACB:

30°	50°	72°	16.3°	81.5°

4. Calculate the approximate number of degrees in one radian, using your calculator.

   You need to divide 360 by 2π. If your calculator has a π button, use it, otherwise type in 3.14 for π.

   If your calculator does not have a π button, this suggests that you are using a very basic calculator. For the later work in this book you will need a scientific, preferably graphical, calculator.

Practical Assignment (1)

Draw a circle with radius 8 cm. Cut a piece of thread exactly 8 cm long. Lay the thread along an arc and mark the two ends of this arc A and B. Mark the centre of the circle O and draw in the radii OA and OB.

Now measure the angle AOB in degrees and write down your answer.

Compare this answer with your answer to question 4 in Practice Questions (1.4).

These two answers should be very close.

Seminar discussion

Why would very precise angle measurement be important in the field of astronomy?

Study tip

Ensure that you know all the definitions and facts presented in this chapter before you move on.

2 Comparing lengths

One-minute overview

In this chapter you will get some practice in working with ratios. Ratio is one of the fundamental concepts in trigonometry, and we illustrate ratio by applying the idea to similar triangles. This entails comparing lengths of sides, which is one of the things you need to do when you start working with trigonometry. We also practise using Pythagoras's Theorem, which is a basic and very important rule about the sides of a right-angled triangle.

Ratio

A ratio is a comparison between two quantities. Suppose Jim earns $400 per week and George earns $800 per week. Then we can express the ratio of Jim's earnings to George's earnings like this:

$$\$400 : \$800$$

and since both earnings are in the same units, i.e. $, we can drop the $ sign and write:

$$400 : 800$$

A fundamental point about ratios is that we are only interested in knowing how large one quantity is in comparison with the other - the actual quantities are not important. So if Jim were earning $4000 per week and George were earning $8000 per week, the relationship between their two amounts of earnings would be the same as if they were earning $400 and $800 respectively (although the result for the two people concerned would be rather different!). We can therefore simplify the ratio 400 : 800 by dividing both numbers by any factor which is common to both numbers. The largest common factor here is 400, so we can simplify in one step to the ratio.

$$1 : 2$$

which only tells us what we knew, namely, that George is earning twice as much as Jim. However, it is vital to grasp this method of comparison in order to be able to work with trigonometric ratios.

Ratios are also expressed sometimes as fractions and sometimes as decimals.

So the ratio 1 : 2 is just the same as the fraction ½ or the decimal 0.5.

When we simplify a ratio from 3 : 6 to 1 : 2, this is the same as cancelling a fraction from 3/6 to ½.

Similar triangles

Remember that in mathematics the word "similar" has a tighter definition than in everyday English.

It means "having the same ratio". So a pair of triangles is similar if the ratio of the lengths of the pairs of corresponding sides is the same for all three sides. We see an example of similar triangles in Fig 2.1:

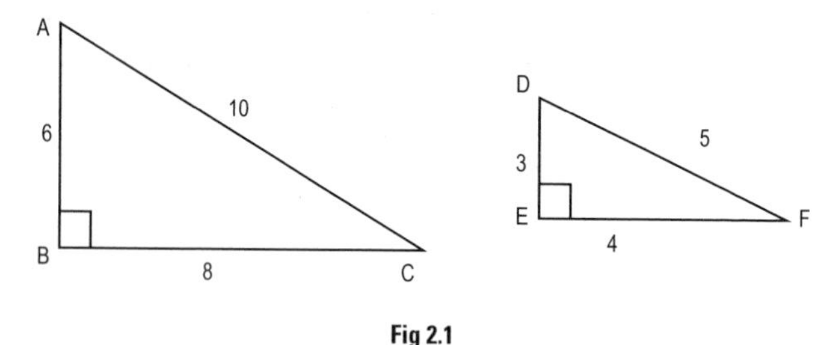

Fig 2.1

Looking at each pair of sides in these two triangles, we see that the sides AB and DE are in the ratio 6 : 3.

We also see that the sides BC and EF are in the ratio 8 : 4 and the sides AC and DF are in the ratio 10 : 5.

The ratio 6 : 3 can be simplified to 2 : 1, as can the ratio 8 : 4 and the ratio 10 : 5.

So the ratios of the lengths of the three pairs of sides in the triangles ABC and DEF are all equal, which is the requirement for two triangles to be (mathematically) similar.

Therefore we can say that triangles ABC and DEF are similar, with sides in the ratio 2 : 1.

A pair of similar triangles will have corresponding pairs of angles equal. You need this fact to answer Practice Questions (2.1). In the example above, the pairs of equal angles are ∠BAC and ∠EDF, ∠ACB and ∠DFE, and the two right angles ∠ABC and ∠DEF.

Practice Questions (2.1)

1. a) Change these ratios to fractions and simplify where possible:
   5 : 15   4 : 20   3 : 7 $16 : $24
   b) Change these ratios to decimals:
   1 : 8   3 : 12   6 : 10 $8 : $100

   It is essential to draw sketches (no measuring needed) for questions 2–4:
2. Triangle PQR has ∠PQR = 60°, QR = 8 cm and QP = 5 cm.
   Triangle STU has ∠STU = 60°, and TU = 12 cm.
   What is the length of ST if the two triangles are similar?
3. Triangle ABC has ∠BAC = 55° and ∠ACB = 65°.
   Triangle DEF is similar to triangle ABC, with ∠EDF = ∠BAC.
   What is the size of ∠DEF?
4. The ratio of similarity between triangles EFG and HKL is 2 : 5.
   EF = 4 cm, FG = 5.6 cm and HL = 8 cm.

> **Note:** the order in which the letters are written shows which sides correspond.
> EF corresponds to HK, FG corresponds to KL and EG corresponds to HL.

Find the lengths of EG, HK, and KL.

Right-angled triangles

A right-angled triangle (see Fig 2.2) has one special feature—an angle of 90°.

But this feature leads to more special mathematical facts about right-angled triangles.

First let's notice some basic points.

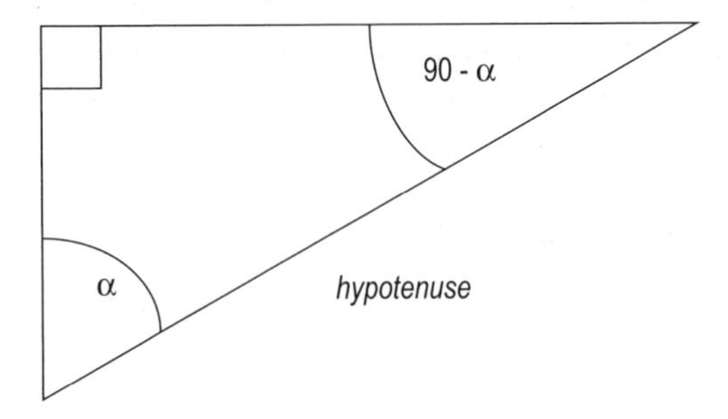

Fig 2.2

The longest side is the side opposite the right angle.
This is called the *hypotenuse.*
As with all triangles, the total of the three angles is 180°.
So the two non-90° angles must add up to 90°.
If one of these angles is α, then the other angle is (90° – α).
The angles α and (90° – α) are complementary angles.

Pythagoras' theorem

This important theorem is not in itself part of the subject of Trigonometry but is very often used along with trigonometrical methods.

Here is a reminder of what this theorem states:

In the triangle in Fig 2.3, h is the length of the hypotenuse.

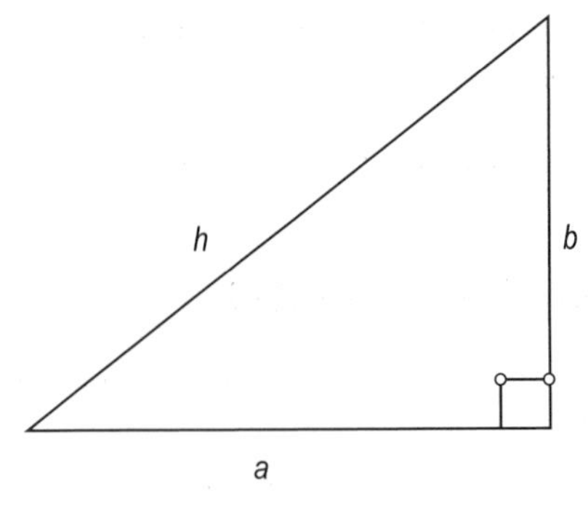

Fig 2.3

We will label the other two sides a and b.

Pythagoras gives us this very important relationship between the lengths h, a and b:

$$h^2 = a^2 + b^2$$

Example (2.1)

In Fig 2.3, let $a = 8$ cm and let $b = 3$ cm.

Then $a^2 + b^2 = 64 + 9 = 73$.

Hence $h^2 = 73$ and $h = \sqrt{73} = 8.54$ cm (correct to 2 d.p.)

Comparing sides of right-angled triangles

The first thing we learn to do in Trigonometry is to work out the ratio of two sides of a right-angled triangle. To do this, we need names for the two short sides of the triangle, as well as the name *hypotenuse* for the long side. Since trigonometric ratios relate to a specific angle, we need to indicate which of the two non-right angles we are referring to. Fig 2.4 shows a fully-labelled right-angled triangle:

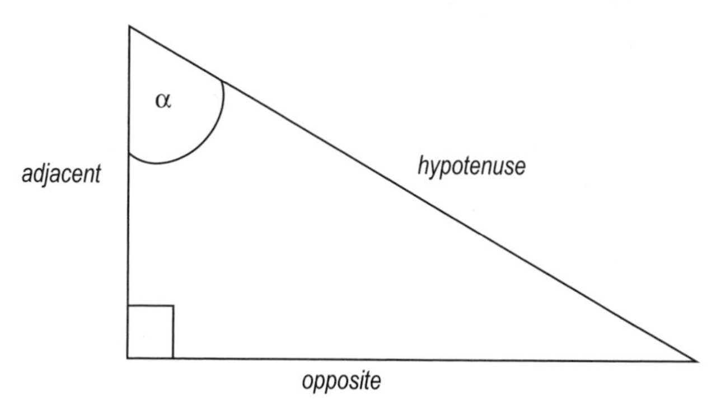

Fig 2.4

The two sides *opposite* and *adjacent* are labelled in relation to the position of the angle we are working with. Let's call this the reference angle. The reference angle in the above diagram is α.

The *adjacent* is always the side between the reference angle and the right angle.

And the *opposite* is the side opposite to the reference angle!

Practice Questions (2.2)

1. A triangle has sides of lengths 4 cm, 6.8 cm and 8 cm. Is it a right-angled triangle?
2. Look at the triangles in Figure 2.1 above. Use Pythagoras's theorem to check that the lengths of the sides shown are consistent with these triangles being right-angled.
3. A right-angled triangle has hypotenuse = 7 cm, and the shortest side is 3 cm.

   Find the length of the other side. Clearly, your answer must be between 3 and 7 cm.
4. Draw a right-angled triangle and label one of the non-right angles α.

   Label the other non-right angle $(90° - \alpha)$ and label the sides of the triangle using $(90° - \alpha)$ as the reference angle. Complete this table by entering Y for a statement which can be true, and N for a statement which cannot be true:

adjacent > hypotenuse	
opposite = adjacent	
hypotenuse = opposite	
$\angle\alpha = 90°$	
opposite < hypotenuse	
opposite + adjacent = hypotenuse	

Tutorial

Progress Questions (2)

1. Draw, as accurately as you can, an isosceles right-angled triangle. Since an isosceles triangle has two equal angles, this means that the two complementary angles are both 45°.

   Measure all the sides accurately and write the results down.

   Taking either angle as the reference angle (since it makes no difference!), write down the ratios opposite : adjacent and opposite : hypotenuse.
2. Use Pythagoras's Theorem to check the accuracy of your measurements of the sides of your triangle in question 1.
3. Draw accurately a triangle ABC with AB = 4 cm. Construct a right angle at B.

   Use a pair of compasses and set the radius to 6 cm. From A, draw an arc cutting the line you have drawn from B. Label the point of intersection C. Measure ∠BAC.
4. Repeat question 3, but this time with AB = 8 cm and the radius of your arc = 12 cm.

   Measure ∠BAC.
5. Draw accurately a triangle DEF with DE = 5 cm, a right angle at E and ∠EDF = 30°.

   Measure the other two sides. With ∠EDF as the reference angle, write down the ratio opposite : hypotenuse.
6. Repeat question 5, but this time with DE any length other than 5 cm.

   If the ratio opposite : hypotenuse is not the same as in question 5, check your work and try again.

Practical Assignment (2)

Sketch a right-angled triangle with one angle very small, say about 10°.

Now sketch another right-angled triangle, this time with the angle even smaller, say about 5°.

What is happening to the opposite side as the angle gets smaller?

What is happening to the complementary angle?

What is happening to the ratio adjacent : hypotenuse?

When the angle reaches 0°, what has happened to the opposite side?

When the angle reaches 0°, what has happened to the ratio adjacent : hypotenuse?

Now look at your triangle with the other angle as reference angle.

What is happening to the ratio opposite : hypotenuse as this angle gets larger?

When this angle reaches 90°, what has happened to the ratio adjacent : hypotenuse?

Seminar discussion

Discuss examples where measurement of triangles might be needed in design, building and other activities.

(A small example: the surface of a wheelchair ramp is the hypotenuse of a right-angled triangle shaped wedge).

Study tip

From the point of view of what follows, the most important part of this chapter has been understanding the names of the sides of the triangle, *adjacent, opposite, and hypotenuse,* and seeing how they compare.

If in doubt, work through the practical assignment again.

3 Understanding trigonometric ratios

One-minute overview

In this chapter we define the three trigonometric ratios, sine, cosine and tangent. We see that for a particular angle, each trigonometric ratio has a unique value. We find out how to work out these values, and how to find the angle if we are given one of its trigonometric ratios. We investigate relationships between the three trigonometric ratios, and we find the values of these ratios for three particularly important angles.

The trigonometric ratio sine

The three ratios sine, cosine and tangent are all of equal importance.

We will look first at **sine.**

Here again is the triangle from Chapter 2.

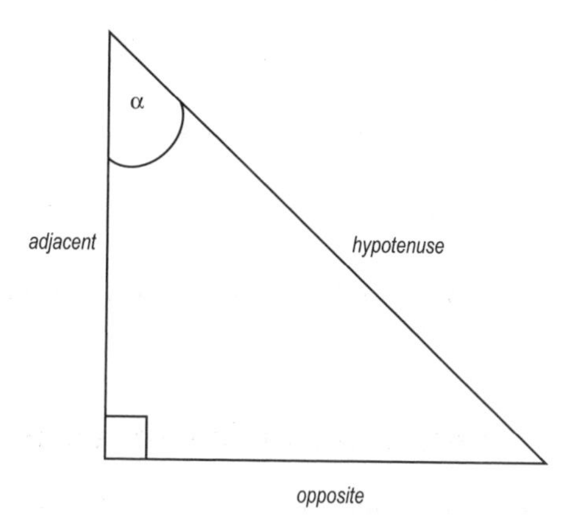

Fig 3.1

The **sine** of $\angle\alpha$ is defined as the ratio of the *opposite* to the *hypotenuse.*

For convenience, from this point onwards we will use the common abbreviation "trig" for "trigonometric". And sine is nearly always written as sin.

Trig ratios are always written as fractions.

So we write:

$$\sin\alpha = \frac{opposite}{hypotenuse}$$

You might have deduced from your work in Progress Questions (2), 5 and 6, that the ratio *opposite : hypotenuse* is fixed for a given angle, even if the angle is drawn in a different size triangle. Indeed, for any given angle, there is one unique value for its sin.

Example (3.1)

In Progress Questions (2), 5, we saw that the ratio *opposite : hypotenuse* for the angle 30° is 1 : 2. Writing this mathematically:

$$\sin 30° = 0.5$$

Practice Questions (3.1)

1. Construct accurately a right-angled triangle with $\angle\alpha = 60°$.

   Measure the opposite and the hypotenuse, and calculate sin 60° to 2 d.p.
2. Construct accurately a right-angled triangle with $\angle\alpha = 42°$.

   Measure the opposite and the hypotenuse, and calculate sin 42° to 2 d.p.

Finding sin ratios on your calculator

Clearly we don't want to have to draw a triangle every time we want to find out the sin of an angle.

Fortunately, we can use the built-in function on a scientific calculator to do this, and the calculator will give us a more accurate result than drawing.

Check that you can use the sin function on your calculator to find the sin of the angles 30°, 60° and 42°.

The values, to 3 d.p, are: sin 30° = 0.500, sin 60° = 0.866, sin 42° = 0.669.

Example (3.2)

One of the ways we can use a trig ratio is to find the size of an angle without drawing and measurement.

We can use sin to find the angle α in this triangle, given that we know the lengths of the opposite and the hypotenuse:

$$\text{We write } \sin\alpha = \frac{opposite}{hypotenuse} = \frac{2.8}{6.4} = 0.4375$$

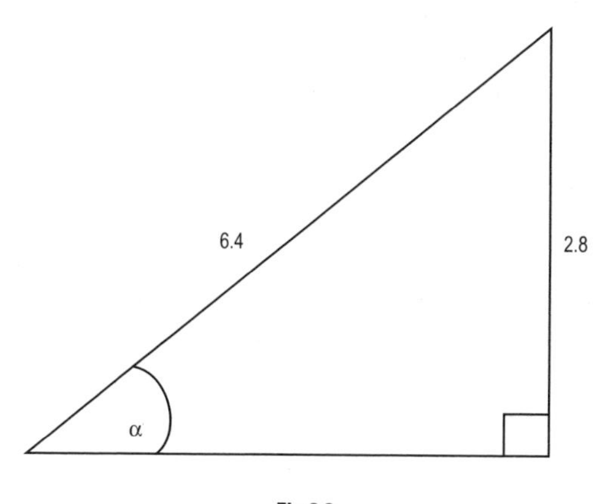

Fig 3.2

Now we want to know the angle for which the sin is 0.4375.

To find this out from your calculator, you will need to use the Inverse or the Shift key (this depends on the type of calculator you have) with the sin function. You may need to check how to do this in your calculator manual.

The angle for which the sin is 0.4375, is 25.9°, correct to 1 d.p.

Terminology

When we find the <u>sin of an angle</u>, we are using the <u>sin function</u>.

When we find the angle for a given sin value, we are using the inverse sin function.

So rather than saying "the angle for which the sin is 0.4375", we can say "inverse sin 0.4375".

The inverse sin function is also sometimes called arcsin. So we can write: arcsin $0.4375 = 25.9°$.

And, arcsin may also be written as $\sin^{-1}$. So we can also write: $\sin^{-1} 0.4375 = 25.9°$.

Caution: The $^{-1}$ in $\sin^{-1}$ is NOT an index. It has nothing to do with $^{-1}$ as used in algebra.

You can use which of these you find easiest, but you must be able to recognise all these terms when you see them.

Practice Questions (3.2)

1. Use your calculator to find, correct to 3 d.p.:
   sin 10° sin 20° sin 30° sin 40° sin 50° sin 60°
   sin 70° sin 80°
2. Using a scale of 1 unit to 10° on the horizontal axis and 1 unit to 0.1 on the vertical axis, plot your results from question 1 and join the points to make a smooth curve.
3. Use your graph in question 2 to estimate:
   arcsin 0.3 arcsin 0.56 arcsin 0.66 arcsin 0.92
   arcsin 0.97
4. Use your calculator to get answers to question 3 to 1.d.p., and compare these with your estimates.

The trigonometric ratios cosine and tangent

The **cosine** of $\angle\alpha$ is defined as the ratio of the *adjacent* to the *hypotenuse*.

We abbreviate **cosine** to **cos**.

So we write:

$$\cos\alpha = \frac{adjacent}{hypotenuse}$$

The **tangent** of $\angle\alpha$ is defined as the ratio of the *opposite* to the *adjacent*.

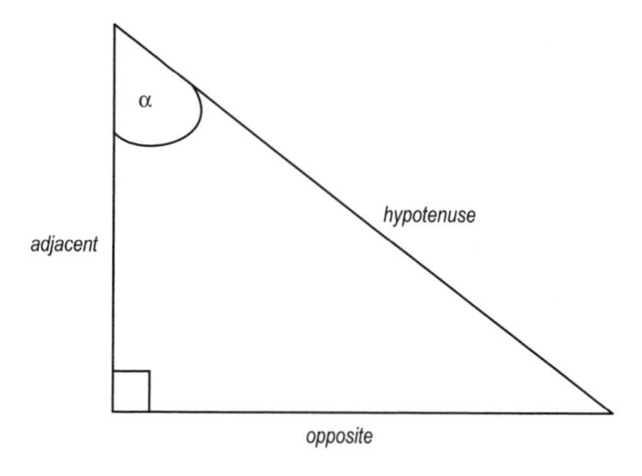

Fig 3.3

We abbreviate **tangent** to **tan.**

> **Note** – that tangent in trigonometry is not connected with the use of the word tangent to indicate a line touching a curve.

So we write:

$$\tan\alpha = \frac{opposite}{adjacent}$$

As with sin, there is one unique value for cos, and one unique value for tan, for a given angle.

Practice Questions (3.3)

1. Measure the adjacent side of your triangle in Practice Questions (3.1), 1, and calculate
   (i) $\cos 60°$ and (ii) $\tan 60°$.
2. Measure the adjacent side of your triangle in Practice Questions (3.1), 2, and calculate
   (i) $\cos 42°$ and (ii) $\tan 42°$.
3. Use the cos and tan functions on your calculator to check your answers to questions 1 and 2.
4. (i) Write down the cos ratio for $\angle\alpha$ in Fig 3.4.
   (ii) Use the inverse cos function on your calculator to find the size of $\angle\alpha$.

5. (i) Use Pythagoras' Theorem to calculate the opposite side in Fig 3.4, correct to 2 d.p. Caution: you need to subtract 3.6^2 from 4.6^2 and then take the square root.
   (ii) Write down the tan ratio for $\angle\alpha$ in Fig 3.4 and use the inverse tan function on your calculator to find the size of $\angle\alpha$.

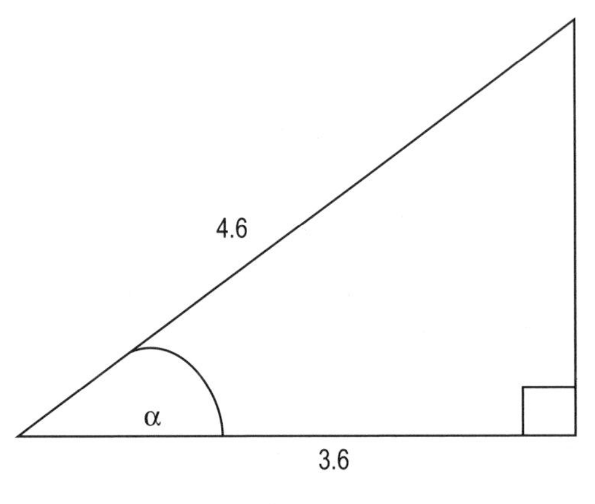

Fig 3.4

Your answer should of course be the same as in question 4, but may differ slightly because of rounding errors in (i).

6. Use your calculator to find, correct to 3 d.p:
   cos 10° cos 20° cos 30° cos 40° cos 50° cos 60° cos 70° cos 80°
7. Using a scale of 1 unit to 10° on the horizontal axis and 1 unit to 0.1 on the vertical axis, plot your results from question 6 and join the points to make a smooth curve.

   Compare this curve with your work from Practice Questions (3.2), 2.
8. Use your graph in question 5 to estimate:
   arccos 0.3 arccos 0.56 arccos 0.66 arccos 0.92 arccos 0.97
9. Use your calculator to get answers to question 8 to 1.d.p., and compare these with your estimates.

Relationships between trig ratios

You may have noticed from your answers to the questions in this chapter that there is a close relationship between sin and cos. The curves you have drawn are the same shape, and in fact the cos curve is a reflection of the sin curve, where the "mirror line" is a vertical line drawn through the point at 45° on the horizontal axis, as shown in Fig 3.5.

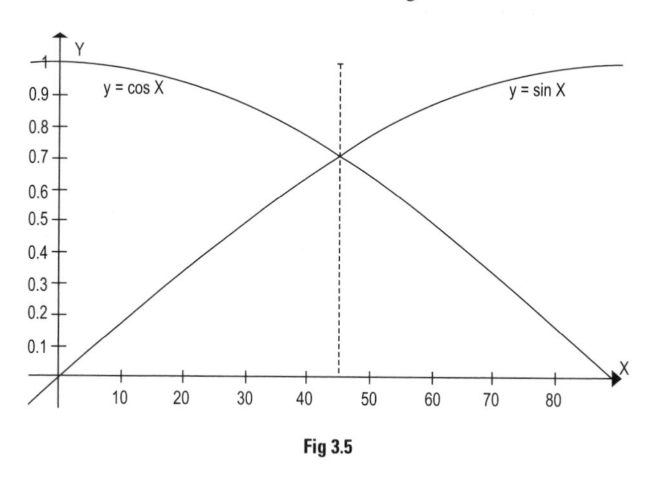

Fig 3.5

We will discover more about these curves in a later chapter, but for now we will establish the exact mathematical relationship between sin and cos.

In Fig 3.6, we can use sin and cos to write down four ratios:

$\sin \alpha = CB/AB$

$\cos \alpha = AC/AB$

$\sin (90 - \alpha) = AC/AB$

$\cos (90 - \alpha) = CB/AB$

Fig 3.6

From these ratios, we can deduce these relationships between a pair of complementary angles:

(i) $\sin \alpha = \cos (90 - \alpha)$ (ii) $\cos \alpha = \sin (90 - \alpha) = AC/AB$

Now look again at Fig 3.6. We can use tan to write down:

$$\tan \alpha = CB/AC \qquad \tan (90 - \alpha) = AC/CB$$

and from these two ratios and the four ratios above, we can deduce these relationships:

(iii) $\tan \alpha = 1/\tan (90 - \alpha)$ (iv) $\tan \alpha = \sin \alpha/\cos \alpha$

Practice Questions (3.4)

1. Use the rules for division of fractions to prove statements (iii) and (iv).
2. Given that sin 15° = 0.259 and cos 15° = 0.966 (values correct to 3 d.p.), find, without using trig functions on your calculator:
   (i) sin 75° (ii) cos 75° (iii) tan 75° (iv) tan 15°
3. Use Fig 3.5 to write down the values of:
   (i) cos 0° (ii) sin 0° (iii) cos 90° (iv) sin 90°
4. Use your calculator to find values of tan α for $\alpha = 10°, 20°$ etc up to 80°.
   Using a scale of 1 unit to 10° on the horizontal axis and 1 unit to 0.5 on the vertical axis, plot these values on a graph. Notice the difference in shape between this curve and the sin and cos curves.
5. Use your graph from question 4 to estimate:
   (i) arctan 0.70 (ii) arctan 1 (iii) arctan 2.1

Three special angles

The angles 30°, 60° and 45° are particularly important. It is very useful to know from memory the trig ratios for these angles.

We have already seen that sin 30° = 0.5 and sin 60° = 0.866 (to 3 d.p.)

We look now at the standard method of finding the trig ratios for these angles.

Fig 3.7 is an equilateral triangle.

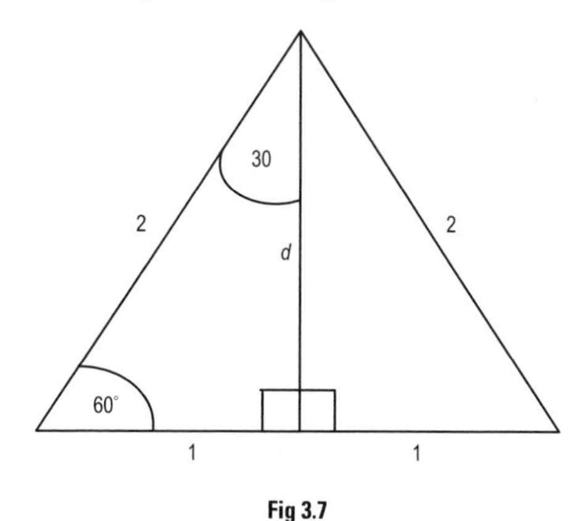

Fig 3.7

Each side is of length 2 units and each angle is 60°. The line of symmetry shown bisects the angle at the top vertex and meets the base at right angles.

So each half of the equilateral triangle is a right angled triangle with angles 30° and 60°, hypotenuse length 2 units and the shortest side length 1 unit.

Using Pythagoras, we can find the length d:

$$2^2 = 1^2 + d^2$$

$$4 = 1 + d^2$$

$$3 = d^2$$

$$d = \sqrt{3}$$

$\sqrt{3}$ is an irrational number and cannot be written as an exact decimal. It is therefore better to work with the length d in this form rather than express it as an inexact decimal.

Now we can use the lengths in the triangle to write down the three trig ratios for the angles 30° and 60°:

$$\sin 30^\circ = \frac{1}{2} \qquad \sin 60^\circ = \frac{\sqrt{3}}{2}$$

$$\cos 30^\circ = \frac{\sqrt{3}}{2} \qquad \cos 60^\circ = \frac{1}{2}$$

$$\tan 30^\circ = \frac{1}{\sqrt{3}} \qquad \tan 60^\circ = \sqrt{3}$$

Finally, we look at the standard method for finding the trig ratios for the angle 45°.

Fig 3.8 is an isosceles right-angled triangle. The two short sides have length 1 unit and the two equal angles are 45°. We can use Pythagoras to find the length of the hypotenuse:

$$h^2 = 1^2 + 1^2$$

$$h^2 = 2$$

$$h = \sqrt{2}$$

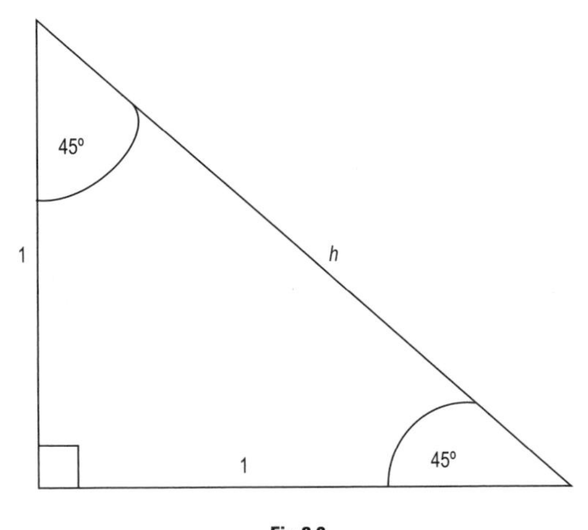

Fig 3.8

Again, $\sqrt{2}$ is an irrational number so we leave it in this form

So here are the trig ratios for the angle 45°:

$$\sin 45^\circ = \frac{1}{\sqrt{2}} \qquad \cos 45^\circ = \frac{1}{\sqrt{2}} \qquad \tan 45^\circ = 1$$

Study all these results, write down the methods for finding them, and learn them from memory.

You will find it very useful later to know these.

Tutorial

Progress Questions (3)

1. Sketch a right-angled triangle with sides of any length. Use your diagram to explain why:
   (i) for any angle α in the range 0° to 90°, sin α and cos α are always in the range 0 to 1.
   (ii) for any angle $\alpha > 45°$ and $< 90°$, tan $\alpha > 1$.
2. Sketch a right-angled triangle with the two shorter sides 3 cm and 6 cm.
   Use the tan ratio to find the two acute angles in the triangle.
3. Draw accurately a right-angled triangle with one angle = 32°. Measure the appropriate sides and find tan 32° from your diagram. Use your calculator to check your answer.
4. Use your calculator to convert the values obtained above for sin 60°, tan 30°, tan 60°, and sin 45°, to decimals correct to 3 d.p.
5. In a right-angled triangle with angles 30°, 60° and the shortest side = 2 units, find the lengths of the other two sides. (Refer to Fig 3.7 if you need help).

Practical Assignment (3)

A warning sign on a road tells you that you are coming to a stretch of road with a gradient 1 in 10.

This means that the vertical height you climb is 1 unit for every 10 units you travel horizontally.

At what angle is the road sloping?

Seminar discussion

In which types of sport might it be important to be able to calculate angles and gradients?

Study tip

Learn the definitions of sin, cos and tan. Many people do this by remembering a mnemonic, such as:

Some **O**ld **H**ag **C**racked **A**ll **H**er **T**eeth **O**n **A**pples (Sin = Opposite / Hypotenuse, Cos = Adjacent / Hypotenuse, Tan = Opposite / Adjacent).

4 How to solve problems

One-minute overview

In this chapter we put into practice all that you have learned about sin, cos and tan. We use these ratios to solve problems of measurement. We represent each problem by a diagram from which we can write down an equation. We then solve the equation to find the length or the angle we are looking for. We extend these techniques to solving problems in 3 dimensions.

Problems using trig ratios in right-angled triangles

Example (4.1)

Joe has a 4 metre ladder. He wants to paint the outside of his bedroom window frame. To reach the window frame, he needs the top of the ladder to reach 3.7 metres up the wall.

Find the angle between the ladder and the ground.

In Fig 4.1, α is the required angle, the ladder is the hypotenuse, and the wall is the opposite side.

As the two known sides are the opposite and the hypotenuse, the ratio we use for this problem is sin.

We write

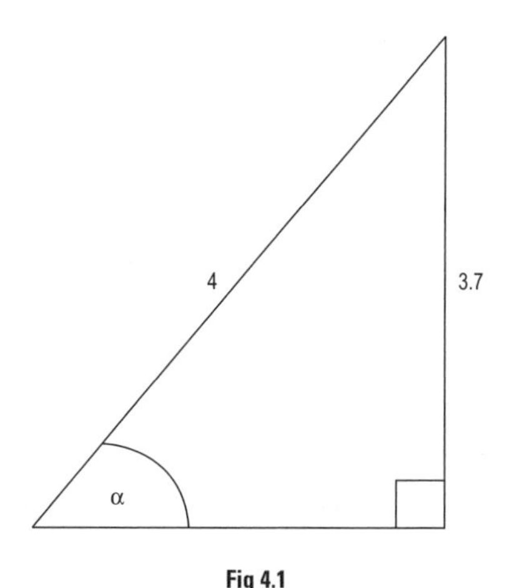

Fig 4.1

$\sin \alpha = 3.7/4$

$\sin \alpha = 0.925$

Now we need the inverse sin function to find α

The calculator gives arcsin $0.925 = 67.7°$

So the angle between the ladder and the ground is 67.7°.

Example (4.2)

A building contractor is required to construct the roof of a house with a minimum slope of 15°.

The distance from the front wall to the back wall of the house is 8 metres. To work out how much roofing material is needed, the builder needs to know the sloping length of the roof.

This problem involves two identical (congruent) triangles as shown in Fig 4.2.

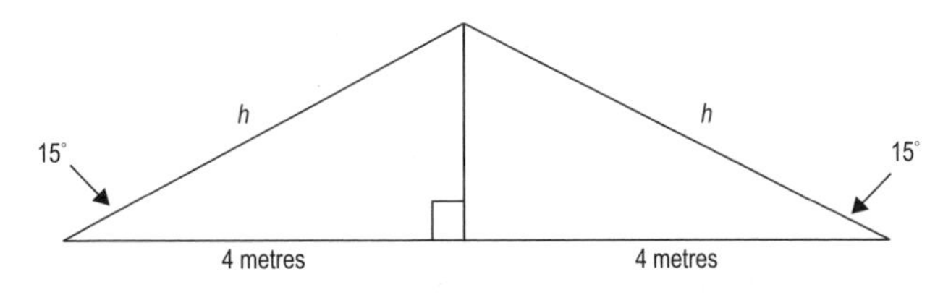

Fig 4.2

The length of the base of each triangle is 4 metres. This is the adjacent side to the 15° angle.

We want to find the sloping length of the roof, which is the hypotenuse. Since we are working with the adjacent and the hypotenuse, the ratio we use here is cos.

So we write

$$\cos 15° = 4/h$$

h is the length we want to find, so we rearrange the equation like this:

$$h = 4/\cos 15°$$

and now we can use the calculator to work out $4 \div \cos 15°$, and we get $h = 4.14$ metres.

Example (4.3)

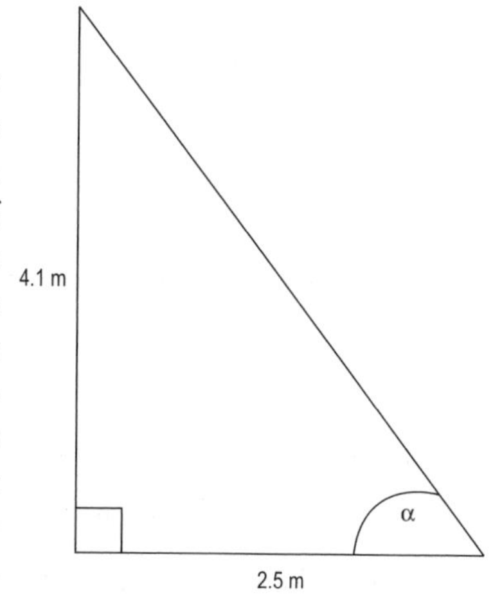

Fig 4.3

Here is another ladder problem! This time Joe needs to work on the front window of a house where the closest he can get his ladder to the wall is 2.5 metres (because of steps down to the basement in front of the house). The window is 4.1 metres above ground level. What is the angle between the ladder and the ground?

The two known sides here are the opposite and the adjacent.

So we use tan:

$$\tan \alpha = 4.1/2.5$$
$$\tan \alpha = 1.64$$
$$\alpha = \arctan 1.64 = 58.6°$$

So the angle between the ladder and the ground is 58.6°.

Example (4.4)

You are standing 100 metres away from a tree. You measure the angle to the top of the tree as 25°.

How tall is the tree?

The height of the tree is marked *h* in Fig 4.4.

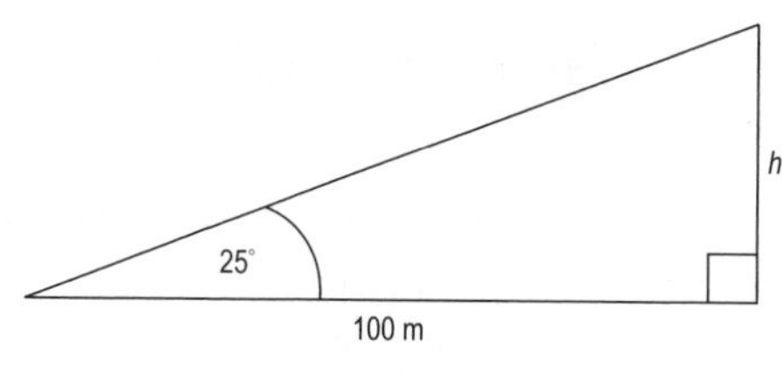

Fig 4.4

The tree is the opposite side to the angle 25°.

The distance to the tree, 100 metres, is the adjacent side.

So in this problem we are again working with the opposite and the adjacent, so the ratio we use is tan.

$$\tan 25° = h/100$$

The calculator gives tan 25° = 0.466, so the equation becomes

$$0.466 = h/100$$

Rearranging the equation gives:

$$h = 100 \times 0.466 = 4.66 \text{ metres.}$$

So the height of the tree is 4.66 metres.

Practice Questions (4.1)

In these questions, you must draw a diagram and label the given lengths or angles. Then identify which two sides of the triangle you are working with, so that you can decide which ratio you need to use.

1. The slope of a wheelchair ramp measures 1.4 metres. The ramp slopes at an angle of 10°. What is the height of the ramp at its highest point?
2. A road has gradient 1 in 20. What is the angle associated with this gradient? Give your answer correct to 2 d.p.
3. A child, who is standing 15 metres away from the foot of a tree, is flying a kite with a string of length 20 metres. The kite gets stuck in the top of the tree. What is the angle between the kite string and the ground? Give your answer correct to 1 d.p.

Angles of elevation, angles of depression

Two terms commonly used in Trigonometry are the "angle of elevation" and the "angle of depression".

The angle of elevation is the angle between the ground and the line of sight when we are looking upwards to an object, as in Example 4.4 and Practice Questions (4.1), 3. The angle of depression is the angle between the horizontal and the line of sight when we are looking downwards to an object.

Example (4.5)

A buzzard is perched on the top of a tree, which is 5 metres high. The buzzard is looking at a mouse, which is on the ground 30 metres away from the tree. (We are assuming that the ground is flat.)

The angle of depression is marked α. This is the angle between the horizontal and the line of sight of the buzzard. The horizontal line drawn from the top of the tree is parallel to the ground; hence the angle of elevation for the mouse looking up at the buzzard is equal to α. (These are alternate angles).

In the right-angled triangle, we can write:

$$\tan \alpha = 5/30$$
$$\tan \alpha = 0.16666\ldots$$
$$\alpha = 9.46^\circ$$

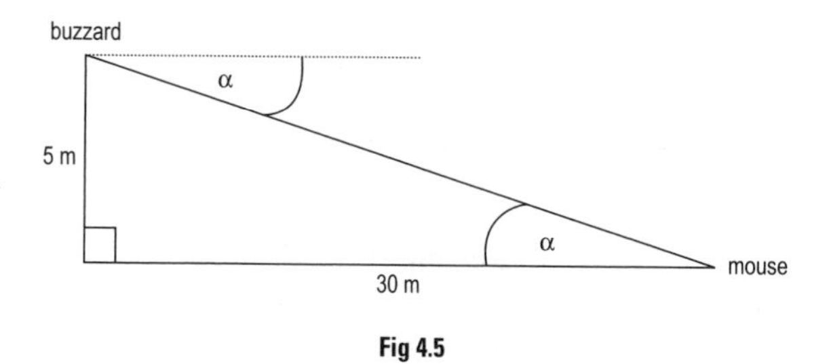

Fig 4.5

Calculator accuracy

In Example 4.5, tan α is a recurring decimal. Very often we get a value for sin, cos or tan which is a recurring decimal or has many decimal places. When this happens, we can get the most accurate answer by keeping the given value in the calculator, and then using the inverse function to find the angle.

In Example 4.5, if we round tan α to 3 d.p, and write $\tan \alpha = 0.167$, and then use the inverse tan function, we get the answer 9.48°. This may seem a small difference from 9.46°, but such errors can make significant differences in longer problems where there are more stages in the working. Always work with more decimal places than are required in the answer, and round off only when you have completed all the calculations.

Practice Questions (4.2)

In all these questions, draw a clearly labelled diagram. Give your answers correct to 1. d.p.

1. A blackbird flies from the top of a hedge, which is 1.6 metres high, and catches a worm that is on the ground. The blackbird flies at an angle of depression of 28°. How far does the blackbird fly?
2. Rudi is watching an aeroplane in the sky. Rudi is at a horizontal distance of 30 kilometres from the aeroplane. The angle of elevation is 19°. What is the height of the aeroplane above the ground?
3. The angle of depression from the top of a cliff to a boat that is 200 metres from the shore is 47°. What is the height of the cliff?

Working in 3 dimensions

We can use trigonometry to solve problems in three dimensions. Using trigonometry in 3 dimensions is just the same as in two dimensions, once we have identified the triangle(s) we are working with.

Here is an example.

Example (4.6)

An explorer comes across a pyramid in the desert. She measures the base and finds that it is square, 100 metres by 100 metres. She measures the angle between one corner edge and the ground as 38°. She wants to work out the height of the apex above the ground.

In the diagram, X is the centre of the square base. The vertex, V of the pyramid is vertically above X. Angle VXC = 90^0.

The "corner edge" is VC. The angle between VC and the ground is angle VCX.

The height the explorer wants to know is VX.

In the right angled triangle VXC we know the angle VCX but we need to know either the length VC or the length XC in order to use one of the trig ratios to find VX.

The explorer could climb the corner edge VC, to measure this length. But she decides it would be easier to use her measurement of BC and AB and apply Pythagoras.

She draws a sketch of the base of the pyramid.

(See Fig 4.7).

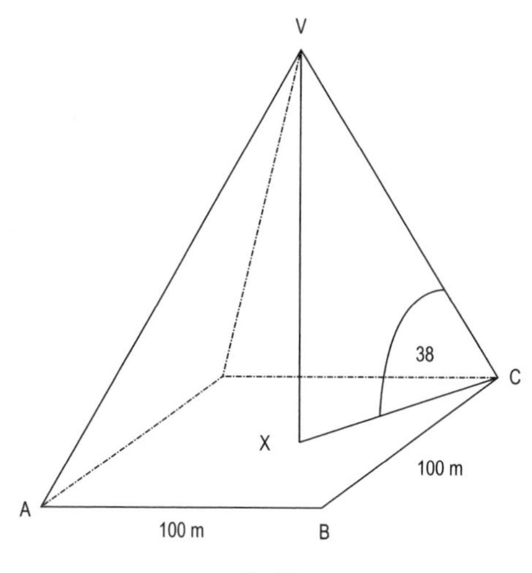

Fig 4.6

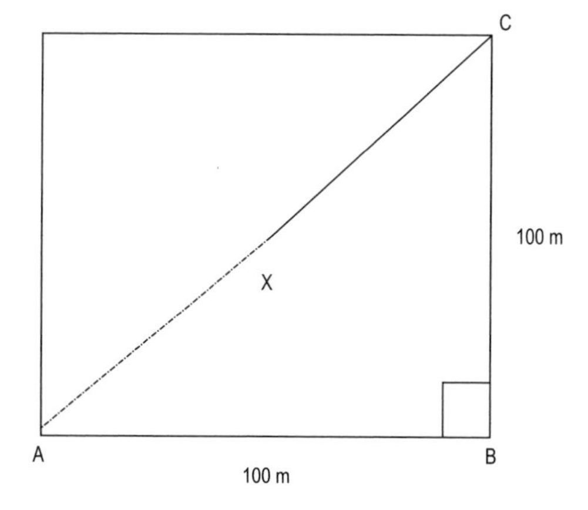

Fig 4.7

X is the centre of the square base.

$$XC = \frac{1}{2} AC$$

We can use Pythagoras to find the length AC:

$$AC^2 = AB^2 + BC^2$$

$$AC^2 = 100^2 + 100^2$$

$$AC^2 = 20000$$

$$AC = \sqrt{20000} = 141.4 \text{ m (to 1 d.p.)}$$

$$XC = \frac{1}{2} AC = 70.7 \text{ m (to 1 d.p.)}$$

So now there is one known length and one known angle in the triangle VXC.

The explorer now draws a sketch of triangle VXC.

In triangle VXC (Fig 4.8), we want to find the height VX.

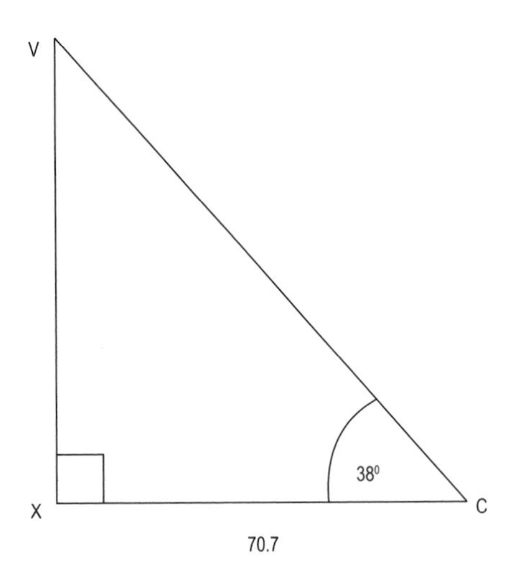

Fig 4.8

$$\tan 38° = VX/70.7$$

$$VX = 70.7 \times \tan 38°$$

$$VX = 55.2 \text{ m (to 1 d.p.)}$$

So the height of the pyramid is 55.2 metres.

Example (4.7)

The explorer travels further and comes across another pyramid. This one has a square base measuring 60 metres by 60 metres, and it appears to be in the same proportions as the previous pyramid. She therefore estimates that the vertical height is 60/100 times 55.2 metres, i.e. 33.1 metres.

The explorer would like to climb up one of the sloping faces of this pyramid, but she would like to know the angle of slope before she attempts the climb. Unfortunately she has lost her measuring instruments so she cannot measure this angle. She will have to use trigonometry to do this.

In Fig 4.9, M is the mid-point of BC.

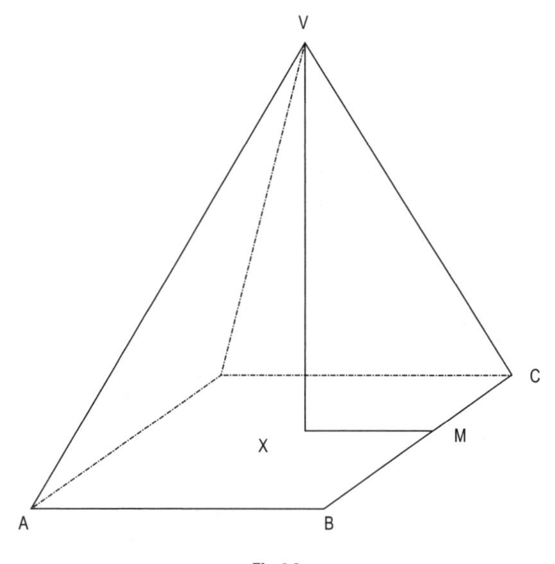

Fig 4.9

The angle between the sloping face and the horizontal is angle VMX.

We need a separate sketch of triangle VMX, and this is shown in Fig 4.10, where angle VMX is marked α.

So, in Fig 4.10, we see:

$$\tan \alpha = 33.1/30$$

$$\alpha = \arctan (33.1/30) = 47.8°.$$

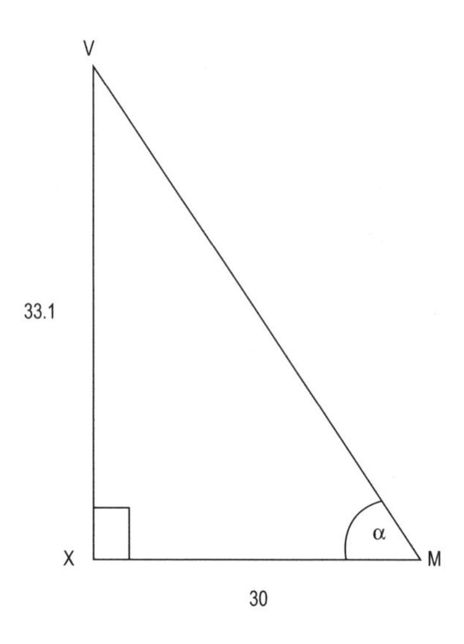

Fig 4.10

You see from the methods shown in these examples that when solving 3-dimensional problems, you need a 2-dimensional diagram at each stage of the working. Don't be tempted to skip this very important step.

Practice Questions (4.3)

1. A pyramid has a square base with side 200 metres. A sloping edge measures 150 metres.

   Find the angle between the sloping edge and the base.

   (Hint: the angle you need is between the sloping edge and the diagonal of the square base).
2. A pyramid has a square base with side 120 metres. The vertical height is 60 metres.
   (i) Find the angle between the sloping edge and the base.
   (ii) Find the angle between a sloping face and the base.

Tutorial

Progress Questions (4)

1. A slide at a children's playpark consists of a sloping section, a flat horizontal section at ground level, and steps at the back. The length of the flat section is 1.25 metres. The top of the slide is 1.5 metres above the ground. The steps slope at 30° to the vertical, and the sloping section of the slide slopes at 40° to the horizontal. Draw a diagram and mark the given lengths and angles, and then calculate the total distance up the steps and down the slide.
2. A new restaurant is to be built beside a busy main road. The restaurant must be set back at least 15 metres from the edge of the road. The approach and departure slip roads must both be at an angle of no more than 20° to the road.
   (i) Draw a diagram showing the main road, the slip roads and the position of the restaurant.
   (ii) Calculate the minimum total length of the two slip roads, correct to 1 d.p.
3. Having built the restaurant, the builder decides to provide a footbridge to make the restaurant accessible from the other side of the road. The angle of elevation of the steps up to each end of the footbridge is to be 55°. The clearance of the bridge above the road is to be 5 metres, and the length of the horizontal section of the bridge is equal to the width of the road. How far from the edge of the road is the base of the steps?
4. A slab of cheese in the shape of a cuboid measures 15 cm by 12 cm by 8 cm. It is to be cut into two equal wedges of width 12 cm, length 15 cm, and height 8 cm. Find the angle of the slope of the wedge.

Practical Assignment (4)

Choose a rectangular shaped room in your house. Measure the height of the room. Measure the distance along the floor from one corner (A) to the opposite corner (B). Find the angle of elevation from corner A to the corner of the ceiling above corner B.

Seminar discussion

How might the methods of trigonometry be useful when designing a gymnasium?

Study tip

You should be confident by now with using trig functions and inverse trig functions on your calculator.
If you have had difficulty with this, then revise this now before you move on.

5 A wider range of angles

One-minute overview

In this chapter we find out what happens to trig ratios for angles larger than 90° and negative angles. We discover that many angles can have the same sin value, cos value or tan value, and we see how the curves we looked at in Chapter 3 can be extended.

Angles outside the range 0° to 90°

In Chapter 1 we looked at angles in the range 0° to 360°. But the range of angles we can work with is not limited to angles within one full turn. For example, the angle –50° is equivalent to the angle 310°, since a clockwise turn through 50° gives the same result as an anticlockwise turn through 310° (or *vice versa*). So every positive angle has a negative equivalent.

What about angles larger than 360°? The angle 370° is equivalent to the angle 10°. In general, for any angle greater than 360° we can subtract multiples of 360° and arrive at an equivalent angle. The angle 800° is equivalent to $800° - (2 \times 360°) = 800° - 720° = 80°$.

Practice Questions (5.1)

1. For each of these angles, find an equivalent angle in the range 0° to 360°:
   400° –400° –180° –90° 1080° –800°
2. Working in radians, for each of these angles, find an equivalent angle in the range 0 to 2π:
   3π 4π $-\frac{1}{2}\pi$ $-\pi$ -5π $-\frac{3}{4}\pi$

Trig ratios for larger angles

Since every angle outside the range 0° to 360° has an equivalent angle inside this range, if we have a method for finding trig ratios for angles in the range 90° to 360°, we will be able to find the trig ratio of any angle.

Fig 5.1 shows a circle divided into four sections. These sections are called quadrants and are numbered 1 to 4 as shown. The lines which divide the circle into quadrants are *x*– and *y*–axes as on a Cartesian graph, and the + and – signs at the ends of the axes show which are the positive and which are the negative directions. The signs of the axes in the four quadrants are:

Quadrant	x	y
1	+	+
2	–	+
3	–	–
4	+	–

First quadrant

The triangle shown in the first quadrant in Fig 5.1 has all three sides positive.

The height (the opposite side to the angle 35°) is positive because the y-direction is positive.

The width (the adjacent side to the angle 35°) is positive because the x-direction is positive.

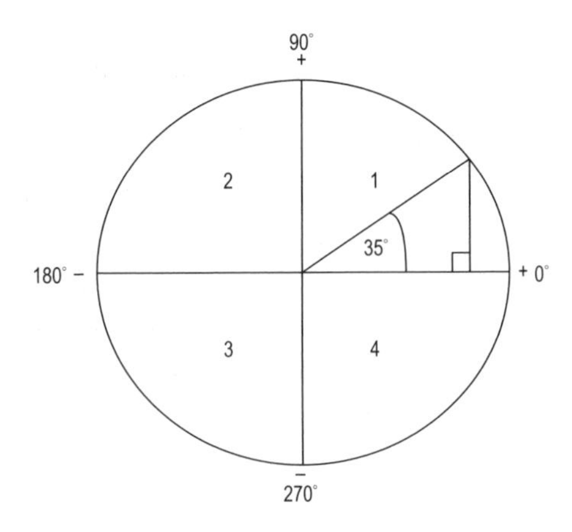

Fig 5.1

The hypotenuse is positive because it is the radius of the circle, which is always positive.

Hence, all three trig ratios are positive in this quadrant. We have worked with angles in the range 0° to 90° and we have seen that the sin, cos and tan values have all been positive.

Second quadrant

Fig 5.2 shows the obtuse angle 145° and its supplementary angle 35°.

We see that in the second quadrant in Fig 5.2 the height of the triangle (the side opposite to the angle 35°) is positive, and the width (the side adjacent to the angle 35°) is negative. The hypotenuse is the radius of the circle and is therefore positive. The signs of the ratios are:

$$\sin = \text{opposite/hypotenuse} = \text{positive/positive} = \textit{positive}$$

$$\cos = \text{adjacent/hypotenuse} = \text{negative/positive} = \textit{negative}$$

$$\tan = \text{opposite/adjacent} = \text{positive/negative} = \textit{negative}$$

Hence, the method for finding the trig ratio of an obtuse angle is to find the supplementary angle, and apply the appropriate sign.

In this example, we have (i) $\sin 145 = \sin 35$ (ii) $\cos 145 = -\cos 35°$ (iii) $\tan 145° = -\tan 35°$.

You can use your calculator to check these results.

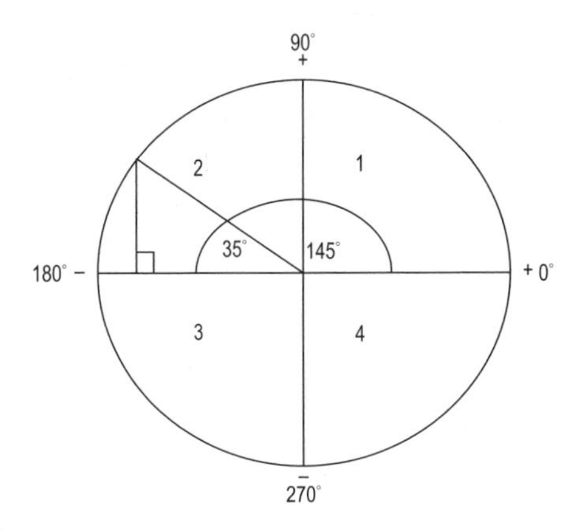

Fig 5.2

Practice Questions (5.2)

1. Find the sin, cos and tan of 215° in relation to the angle 35°, using a similar method to that shown above. Draw a diagram similar to Fig 5.1 and Fig 5.2, but this time with the angle 215° measured anticlockwise from the x-axis in the first quadrant. You will have an angle of 35° in the third quadrant, with the adjacent side of the triangle on the negative x-axis. This triangle will be a reflection in the x-axis of the triangle shown in Fig 5.2. Now work out the signs of the trig ratios in the third quadrant.
2. Find the sin, cos and tan of 325° in relation to the angle 35°, using a similar method in the fourth quadrant. The triangle you construct should be a reflection in the x-axis of the triangle shown in Fig 5.1.
3. Complete this table:

Quadrant	sin	cos	tan
1	+	+	+
2	+	–	–
3			
4			

4. Write down the sign (+ or –) for these trig ratios, then use your calculator to check your answers:
   sin 100° cos 100° tan 160° sin 280° cos 280°
   sin (–20°) cos (–20°) tan (–100°)

Summarising the method

Although we can use a calculator to find trig ratios for any angle, it is very important to get completely familiar with the signs of the trig ratios in each quadrant. With this knowledge, many trig problems, for example trig equations, become much easier. It is therefore a good idea to assume for the moment that your calculator can only give you the trig ratios for acute angles.

Then, to find the trig ratios for angles larger than 90°, these are the steps we take:

(i) Obtuse angles (over 90° and up to 180°)
Find the supplementary angle, i.e. subtract the acute angle from 180°.

Find the trig ratio of the supplementary angle.
For cos or tan, make this value negative.

(ii) Reflex angles over 180° and up to 270°:
Subtract 180° from the angle.
Find the trig ratio of the resulting angle.
For cos or sin, make this value negative.

(iii) Reflex angles over 270° and up to 360°:
Subtract the angle from 360°.
Find the trig ratio of the resulting angle.
For sin or tan, make this value negative.

Practice Questions (5.3)

1. Referring back to Chapter 3 if you need to, complete this table (with fraction, not decimal values):

	30°	45°	60°
sin			
cos			
tan			

Now use these answers, and the table from Practice Questions (5.2), 3, to complete the following table with fraction values:

	120°	135°	150°
sin			
cos			
tan			

2. Using your calculator for acute angles only, find the following trig ratios:

sin 170° cos 170° tan 170° cos 205° sin 255° tan 214°

tan 280° cos 290° sin (–20°) cos (–35°) tan 375° sin 500°

Extending the graphs of sin, cos and tan

In Chapter 3 we looked at the graphs of the trig ratios for angles between 0° and 90°.

We now look at what happens to these graphs outside this range of angles.

Let's look first at the sin graph. We will extend the graph, plotting sin values at 10° intervals.

In Chapter 3 we plotted sin values from 10° to 80°.

What is the value of sin 90° ? Your calculator shows that sin 90° = 1.

To understand this result, imagine a right-angled triangle with angles 89° and 1°.

You can see that for the angle 89°, the opposite and the hypotenuse are almost equal.

Now imagine the angle 89° getting larger. When it reaches 90° the triangle disappears.

The opposite and the hypotenuse coincide and are therefore equal in length. Hence sin 90° = 1.

Also note that the adjacent side has disappeared! So the length of the adjacent side is zero, and hence cos 90° = 0.

Note – that we cannot find a value for tan 90°, since we cannot divide the opposite by the adjacent—division by zero is not possible. So tan 90° is undefined.

Then, for angles over 90°, using the method for the second quadrant (see Fig 5.2), we have:

sin 100° = sin 80°, sin 110° = sin 70°, sin 120° = sin 60° etc. up to sin 170° = sin 10°, sin 180° = sin 0°.

Fig 5.3 shows the sin curve in the range 0° to 180°, and we see that it is symmetrical about the line x = 90°:

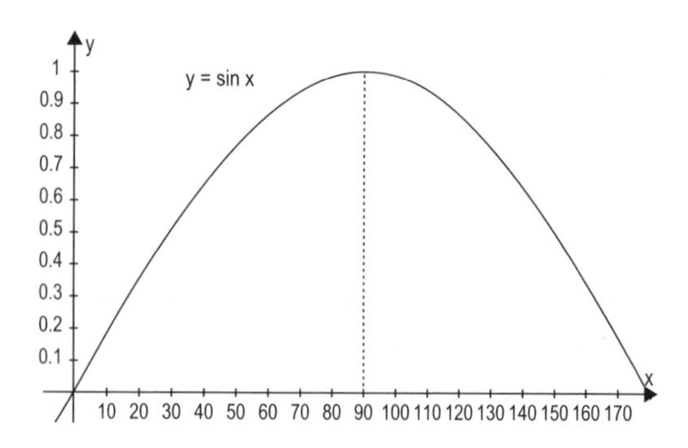

Fig 5.3

Working in the same way for cos gives us the graph in Fig 5.4:

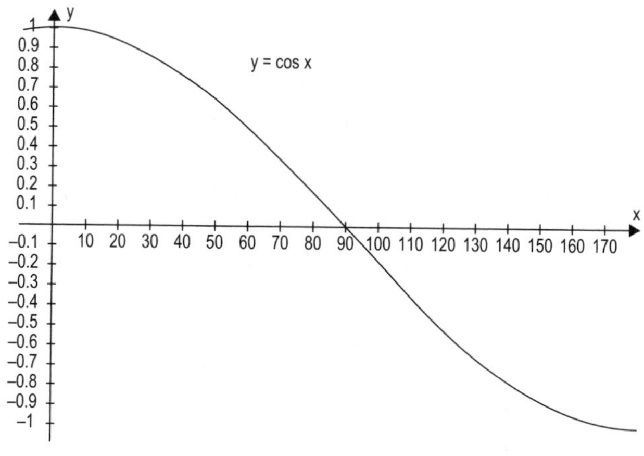

Fig 5.4

and for tan, we get the graph in Fig 5.5:

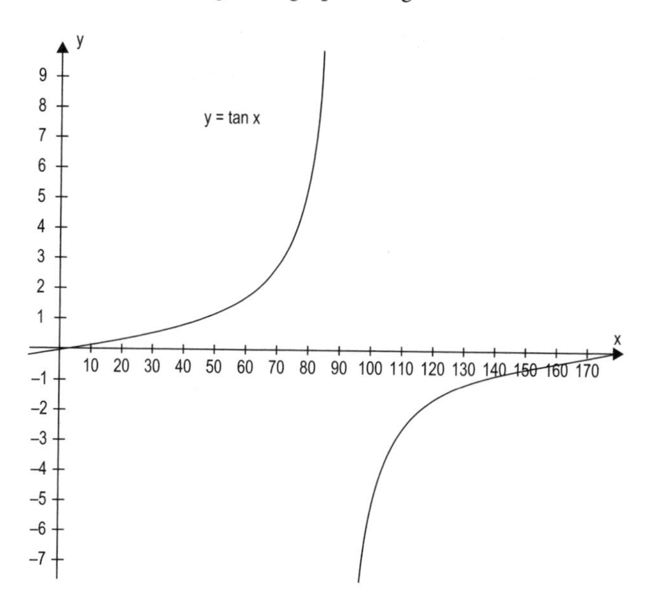

Fig 5.5

Practice Questions (5.4)

1. Using the methods for finding trig ratios in the third and fourth quadrant, sketch the sin, cos and tan graphs in the range 0° to 360°.
2. Use your graphs to answer these questions:
   (i) How many angles in the range 0° to 360° have the same sin value?
   (ii) How many angles in the range 0° to 360° have the same cos value?
   (iii) How many angles in the range 0° to 360° have the same tan value?
3. Write down the values of:
   (i) sin 0° sin 90° sin 180° sin 270° sin 360°
   (ii) cos 0° cos 90° cos 180° cos 270° cos 360°
   (iii) tan 0° tan 180° tan 360°
4. a) Using your calculator, find arcsin 0.4 correct to 1 d.p.
   Your calculator gives you an angle in the range 0° to 90°.
   Now write down a second value for arcsin 0.4.
   (This means look for a second angle whose sin is 0.4; this angle will be in the range 90° to 180°.)
   b) Using your calculator, find arctan 2.58 correct to 1 d.p.
   Your calculator gives you an angle in the range 0° to 90°.
   Now write down a second value for arctan 2.58.
   (This means look for a second angle whose tan is 2.58; this angle will be in the range 180° to 270°.)
   c) Using your calculator, find arccos 0.75 correct to 1 d.p.
   Your calculator gives you an angle in the range 0° to 90°.
   Now write down a second value for arccos 0.75.
   (This means look for a second angle whose cos is 0.75; this angle will be in the range 270° to 360°.)

Tutorial

Progress Questions (5)

1. Using your calculator only in the range 0° to 90°, find the following trig ratios correct to 3 d.p.:
   a) sin 160°, sin 295°, sin 330°, sin (–20°)
   b) cos 100°, cos (–30°), cos 200°, cos 385°
   c) tan 125°, tan 225°, tan (–60°), tan 400°
2. Set your calculator to work in radians, and find arccos 0.8, correct to 2. d.p. Now find two more angles with the same cos value (one of these angles must be outside the range 0 to 2π).
3. The angle $(180° – \alpha)$ has the same sin value as angle α.
   We can state this more neatly like this: $\sin \alpha = \sin (180° – \alpha)$.
   Write down the correct statement for (i) $\tan \alpha$ (ii) $\cos \alpha$.
4. Sketch the graph of $\sin \theta$ in the range –180° to 180°. Choose a value α in the range 0° to 180°.
   What is the relationship between $\sin \alpha$ and $\sin(-\alpha)$?
5. Sketch the graph of $\cos \theta$ in the range –180° to 180°. Choose a value α in the range 0° to 180°.
   What is the relationship between $\cos \alpha$ and $\cos(-\alpha)$?

Practical Assignment (5)

Plot on a graph the times of high tide, or the times of sunset, over a long enough period to see a shape in the curve. Comment on your results.

Seminar discussion

Why are the sin and cos curves the same shape, while the tan curve is completely different?

Study tips

(i) The essence of this chapter is the discovery of the signs of the trig ratios in the four quadrants.
Memorise the results from Practice Questions (5.2), 3. The following diagram may help you.
It shows which ratios are positive in each quadrant.

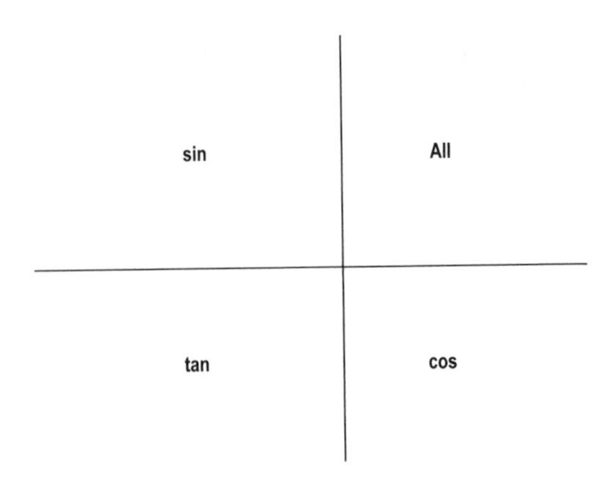

Fig 5.6

(ii) Learn to sketch, from memory, the sin, cos and tan graphs in the range 0° to 360°. We will be returning to these graphs in Chapter 8 and you will need to know their shapes.

6 How to solve problems without right angles

One-minute overview

In this chapter we meet three new formulae which use trig ratios, and we apply these formulae to problems involving non-right-angled triangles, to find angles, lengths and areas. These measurement methods will also be needed in the next Chapter, when we look at bearings, so by working with these methods now, you will be preparing yourself for the next set of problems.

Triangles without right angles

We first defined trig ratios by referring to the three sides of a right-angled triangle.

But we have seen that we can extend trig ratios to angles outside the range 0° to 90°.

And now we see that we can use trig ratios to solve measurement problems in non right-angled triangles.

Labelling convention

It is common practice to label the corners of a triangle with the letters A, B and C, and then to label the sides opposite each corner *a*, *b* and *c*, as shown in Fig 6.1. We use the letters A, B and C to represent the angles at their respective corners.

This system can also be used with other sets of letters, e.g. DEF, PQR etc.

We see that a longer side is opposite a larger angle, and a shorter side is opposite a smaller angle.

There is a mathematical rule which uses this fact, and it is very useful for solving problems involving lengths and angles in triangles.

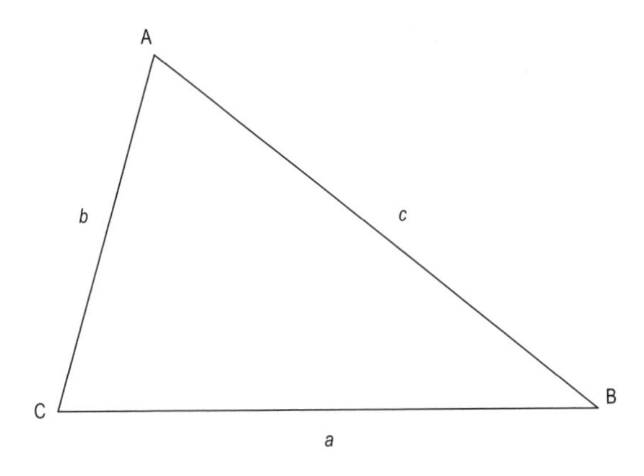

Fig 6.1

The sine rule

In the triangle in Fig 6.2, a perpendicular line has been drawn from A to the side BC.

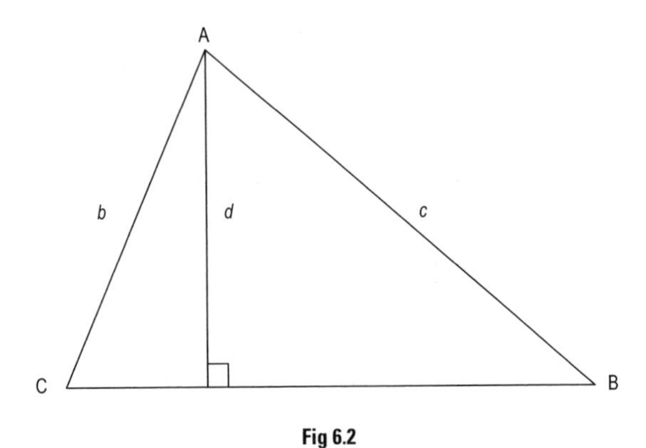

Fig 6.2

So we have two right–angled triangles and we can write:

(1) $\sin C = d/b$ and (2) $\sin B = d/c$
(1) can be rearranged as: $d = b \sin C$
(2) can be rearranged as: $d = c \sin B$

So we see from these two equations that

$$b \sin C = c \sin B$$

which can be re-arranged as

$$\frac{b}{\sin B} = \frac{c}{\sin C}$$

We could also draw a perpendicular from B to side AC, or from C to side AB, and this would enable us to extend the rule to side a and angle A. So the complete rule is:

$$\frac{a}{\sin A} = \frac{b}{\sin B} = \frac{c}{\sin C}$$

We can also write this the “other way up”:

$$\frac{\sin A}{a} = \frac{\sin B}{b} = \frac{\sin C}{c}$$

and in practice we can use the rule in whichever form works better for each particular problem.

This rule is called the sine rule, and you should memorise it now.

Example (6.1)

Fig 6.3 shows a triangle in which two angles and one side are known.

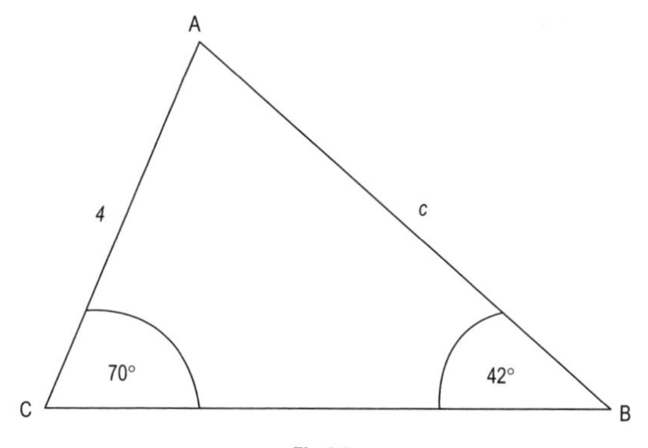

Fig 6.3

With the given measurements, we can use the sine rule to find the remaining sides:

$$\frac{4}{\sin 42°} = \frac{c}{\sin 70°}$$

Re-arranging,

$$c = \frac{4 \times \sin 70°}{\sin 42°}$$

and the calculator gives $c = 5.62$ to 2 d.p.

Then, $\angle A = 180° - (70° + 42°) = 68°$

Applying the sine rule a second time,

$$\frac{4}{\sin 42°} = \frac{a}{\sin 68°}$$

which gives $a = 5.54$ to 2 d.p.

When can the sine rule be used?

We can use the sine rule if we know:

(i) two angles and one side, or
(ii) two sides and an angle which isn't between these two sides.

We have seen case (i) in Example 6.1.

In case (ii) we will see that there are two possible triangles which fit the given data.

Example (6.2)

Fig 6.4 shows two triangles, each with $a = 4.5$, $c = 3.8$ $\angle C = 53°$.

What is $\angle A$? Using the sine rule, we have, in either case:

$$\frac{3.8}{\sin 53°} = \frac{4.5}{\sin A}$$

Re-arranging,

$$\sin A = \frac{4.5 \times \sin 53°}{3.8}$$

The calculator gives $\sin A = 0.9457\ldots$

Inverse sin function = arcsin

and the inverse sin function gives $\angle A = 71°$

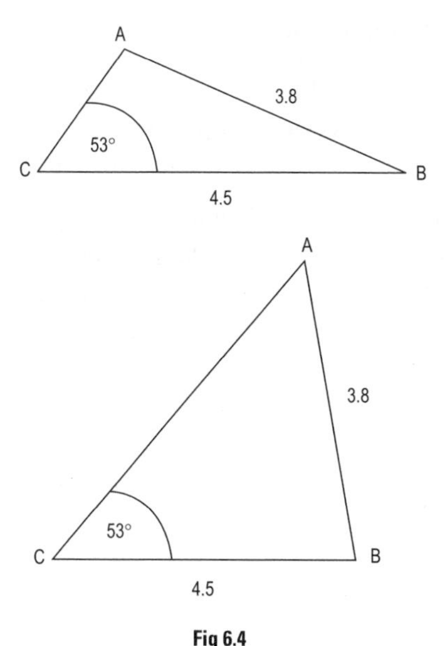

Fig 6.4

But we know that for every acute angle, there is an obtuse angle that has the same sin value.

Remember the rule from Chapter 5: $\sin \alpha = \sin (180° - \alpha)$.

So in this case, $\angle A = 71°$ or $109°$ (These answers are correct to 1 d.p.).

The upper triangle in Fig 6.4 is obtuse-angled, so $\angle A = 109°$, and $\angle B = 180° - (109° + 53°) = 18°$.

In the lower triangle $\angle A = 71°$ and $\angle B = 180° - (71° + 53°) = 56°$.

We can also calculate the length b(AC) which will be shorter in the upper triangle and longer in the lower triangle. (See Practice Questions 6.1, 1) below.

This is a very important result. You must be aware that when you are given two sides and an angle that is not between these two sides, then you have to look for both the acute and the obtuse angle. Whether there are actually two possible triangles depends on the given lengths and angles. You can always get two possible values for the first angle you find, but when you try to calculate the third angle, in some cases, using one of the two angles gives a negative result, and if this happens then there is only one possible triangle.

Practice Questions (6.1)

In each of these questions, sketch the triangle and and label it before you write down the equations.

Give your answers correct to a sensible degree of accuracy.

1. Find the length b in each of the two triangles in Fig 6.4.
2. $a = 10.8$ cm, $\angle A = 49°$, $\angle B = 28°$. Find the lengths b and c.
3. $\angle C = 93°$, $\angle B = 32°$, $b = 6$ cm. Find the lengths a and c.
4. $\angle A = 53°$, $a = 3.5$ cm, $c = 4.1$ cm. Find $\angle C$, $\angle B$ and length b.
5. $\angle C = 48°$, $c = 5.8$ cm, $b = 6.8$ cm. Find $\angle A$, $\angle B$ and length a.
6. $\angle A = 110°$, $a = 12.5$ cm, $b = 8.7$ cm. Find $\angle B$, $\angle C$ and length c.

The cosine rule

The sine rule will not help us to calculate the remaining angles and sides of a triangle if we know

(i) two sides and the angle between the two sides
(ii) all three sides but no angles.

Look at the two triangles in Fig 6.5.

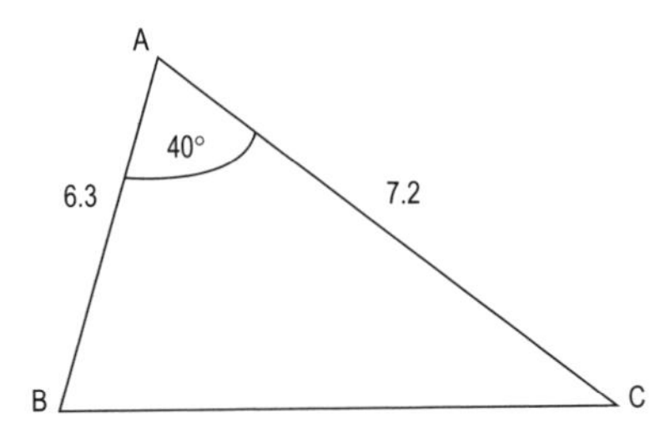

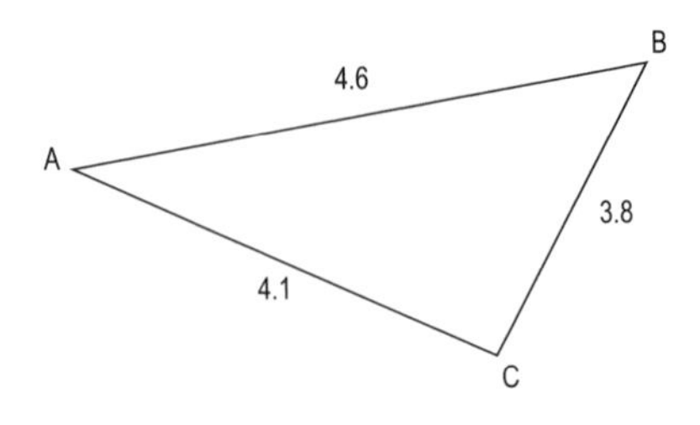

Fig 6.5

The sine rule does not help us with either of these triangles because we cannot pair any side with its opposite angle, so we cannot find the missing lengths and angles.

In these situations we need the cosine rule.

Fig 6.6 below shows a triangle ABC with a perpendicular drawn from B to AC.

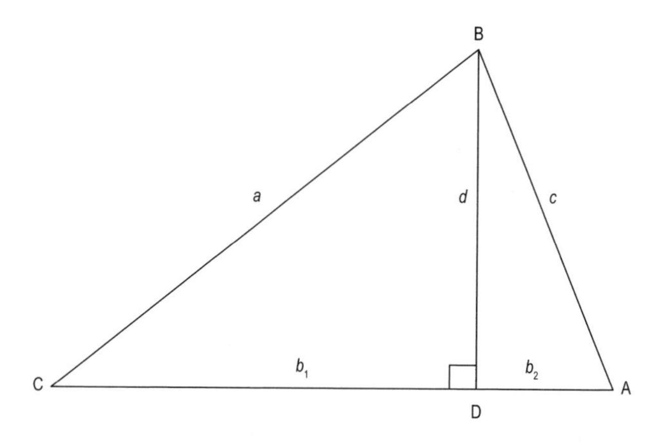

Fig 6.6

The side AC is cut into two sections, so $b_1 + b_2 = b$.

The cosine rule is derived from Pythagoras' Theorem.

In triangle BCD we see that

$$a^2 = d^2 + b_1^{\,2}$$

and in triangle BDA,

$$c^2 = d^2 + b_2^{\,2}$$

Eliminating d from these two equations gives

$$a^2 - b_1^{\,2} = c^2 - b_2^{\,2}$$

Rearranging:

$$a^2 = b_1^{\,2} - b_2^{\,2} + c^2$$

Factorising the first two terms on the right:

$$a^2 = (b_1 - b_2)(b_1 + b_2) + c^2$$

$b_1 + b_2 = b$ so we have:

$$a^2 = b(b_1 - b_2) + c^2$$

$$b_1 - b_2 = b_1 + b_2 - 2b_2 = b - 2b_2$$

$$a^2 = b(b - 2b_2) + c^2$$

Multiply out the bracket

$$a^2 = b^2 - 2b_2b + c^2$$

Now look at triangle BDA, and we see that cos A = b_2/c and we can re-arrange this as $b_2 = c \cos \text{A}$

So the equation becomes

$$a^2 = b^2 - 2bc \cos \text{A} + c^2$$

or

$$\underline{a^2 = b^2 + c^2 - 2bc \cos \text{A}}$$

Note – Unless you are studying for an exam where you might be asked to prove the sine or cosine rule, you do not need to learn the proofs. But you certainly do need to learn the rules.

Look again at Fig 6.6. You can see that we could also draw a perpendicular from A to BC, or from C to AB, and if we had done this we would have derived the cosine rule by the same method but with these results:

$$\underline{b^2 = a^2 + c^2 - 2ac \cos \text{B}} \quad \text{or} \quad \underline{c^2 = a^2 + b^2 - 2ab \cos \text{C}}$$

We can also rearrange the formula for the case when we want to find an angle:

$$\cos \text{A} = \frac{b^2 + c^2 - a^2}{2bc} \quad \text{or} \quad \cos \text{B} = \frac{a^2 + c^2 - b^2}{2ac}$$

$$\text{or} \quad \cos \text{C} = \frac{b^2 + a^2 - c^2}{2ab}$$

Example (6.3)

Fig 6.7 shows a triangle with ∠B = 50°, a = 6 cm and c = 7.2 cm.

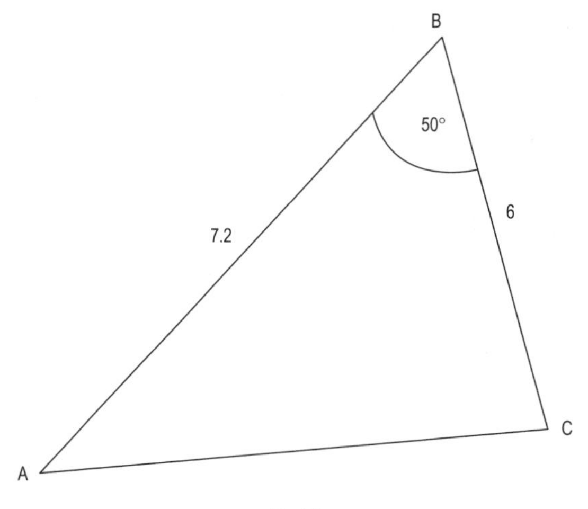

Fig 6.7

We can use the cosine rule to find b.

We need this form of the cosine rule:

$$b^2 = a^2 + c^2 - 2ac \cos B$$

Putting in the given values, we get:

$$b^2 = 6^2 + 7.2^2 - 2(6)(7.2) \cos 50°$$

Do take care to note the order in which the calculation must be done.

The term "$2ac$ cos B" must be calculated <u>before</u> the subtraction is done.

> Doing this type of calculation in the wrong order is a common error in using the cosine rule. Make sure you avoid this error. Check your answers carefully when you work through Practice Questions (6.2).

$$b^2 = 36 + 51.84 - (86.4 \times \cos 50°)$$

$$b^2 = 87.84 - 55.54 = 32.30$$

$$b = 5.68 \text{ cm correct to 2 d.p.}$$

(If we want to find $\angle A$ and $\angle C$, we can now use the sine rule.)

Now we'll use the cosine rule to find an angle where all three sides of the triangle are known.

Example (6.4)

We can use the cosine rule to find any one of the angles in Fig 6.8. Let's choose $\angle A$.

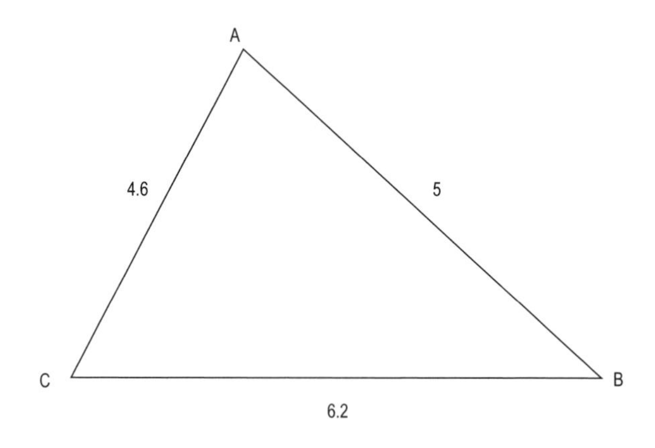

Fig 6.8

To find $\angle A$, we use the cosine rule in this form:

$$\cos A = \frac{b^2 + c^2 - a^2}{2bc}$$

Putting in the values of *a*, *b* and *c*, we have:

$$\cos A = \frac{4.6^2 + 5^2 - 6.2^2}{2(4.6)(5)} = \frac{7.72}{46}$$

which gives $\angle A = 80.3°$.

Note that in this example, the top line of the formula, $b^2 + c^2 - a^2$, gives a positive result.

cos A is positive, and ∠A is an acute angle. If the top line of the formula gives a negative result, then the angle is obtuse.

We could now repeat the use of the cosine rule to find the ∠B or ∠C. But as the sine rule is easier to use than the cosine rule, now that we know one angle, we should use the sine rule to find the other angles.

Practice Questions (6.2)

Give all your answers correct to 1 d.p.

1. Read Example 6.3 again and then find ∠A and ∠C.
2. Read Example 6.4 again and then find ∠B and ∠C.
3. Sketch a triangle DEF with ∠D = 55°, DE = 4 cm and DF = 4.2 cm.

   Label the sides *d*, *e* and *f*. Find the length *d*.
4. Sketch a triangle with sides of lengths 11.0, 12.3 and 16.8 cm.

   Label the angles and sides correctly, and find all the angles.

Note – that the angle opposite the longest side is an obtuse angle in question 4.

Area of a triangle

A commonly-used formula for the area of a triangle is:

$$\text{area} = \tfrac{1}{2}\,\text{base} \times \text{height}$$

In this formula, the height must be the perpendicular height, which means we have to know the perpendicular distance from one corner of the triangle to the opposite side.

In Fig 6.9, h is the perpendicular distance from B to AC, and the area is $\frac{1}{2}bh$.

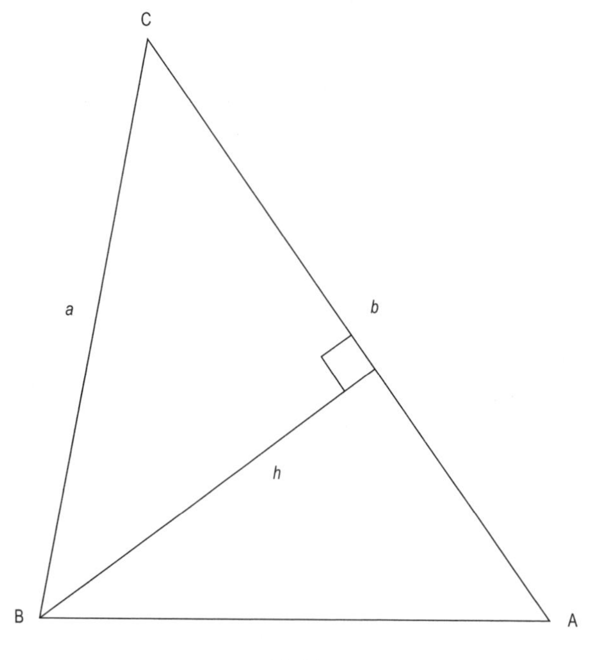

Fig 6.9

The perpendicular cuts triangle ABC into two right-angled triangles, and in the upper triangle we have:

$$\sin C = h/a$$

which can be re-arranged as:

$$h = a \sin C$$

Hence the area of the triangle can be written:

$$\tfrac{1}{2}b(a \sin C) \text{ or } \underline{\tfrac{1}{2}ab \sin C}.$$

The formula can also be used in these forms:

$$\underline{\tfrac{1}{2}bc \sin A}, \text{ or } \underline{\tfrac{1}{2}ac \sin B},$$

and it works in any triangle where we know two sides and the angle between them.

Example (6.5)

In Fig 6.10, we are given the lengths of two sides of the triangle, and the angle between these two sides.

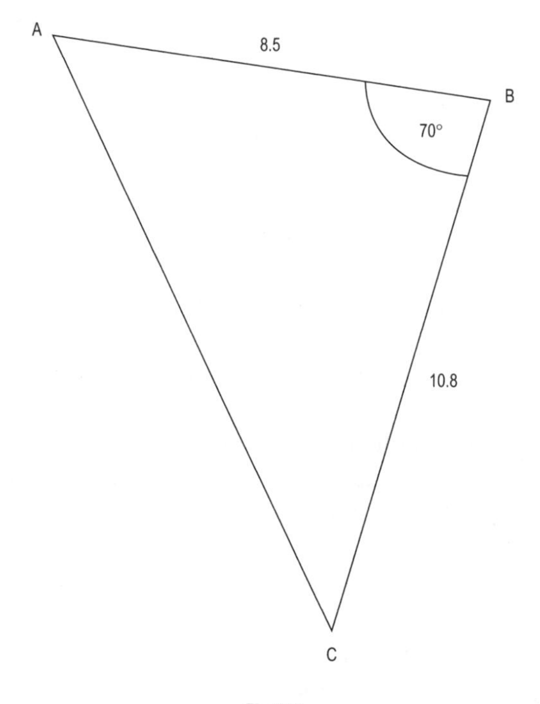

Fig 6.10

The formula ½ *ac* sin B in this case , gives:

$$\text{Area} = \tfrac{1}{2} \times 8.5 \times 10.8 \times \sin 70^\circ$$

$$\text{Area} = 43.1 \text{ square units.}$$

Questions using this formula are included in Progress Questions (6) below.

Tutorial

Progress Questions (6)

1. Sketch a triangle ABC in which $\angle A = 30°$, $\angle C = 68°$ and $AB = 7$ cm.
   (i) Find the lengths of AC and BC.
   (ii) Find the area of the triangle.
2. Sketch a triangle DEF in which $DE = 3.8$ cm, $EF = 5.1$ cm and $FD = 7.4$ cm.
   (i) Find the size of $\angle D$ amd $\angle E$.
   (ii) Find the area of the triangle.
3. Sketch a triangle PQR in which $PQ = 4$ cm, $PR = 7$ cm, and $\angle Q = 80°$.

   Find the length of QR. (You will need to find $\angle R$ and $\angle P$ first).
4. A cat leaves its home and walks 5 km due North. It then turns to the NE (i.e. it turns through an angle of 45° to the right) and walks another 5 km. Draw a sketch of the cat's walk.

   Calculate the distance the cat now has to walk, in a straight line, to get home.
5. Ivan wants to paint a window which is 3 metres above the ground. He has a ladder which is 3.5 metres long, but the ground slopes downwards away from the bottom of the wall at an angle of 5°, and there is a flower bed, which means that the closest he can get his ladder to the base of the wall is 1.5 metres. Draw a diagram and work out whether Ivan's ladder is long enough to enable him to reach the window.
6. Maria wants to tile her floor. The room measures 2 metres by 1.5 metres.

   Maria has a budget of $100 to spend on tiles.

   Each tile is a regular hexagon with side 10 cm as shown in Fig 6.11. Each tile costs $1.

   Can Maria tile her floor within her budget?

   (You can ignore the problem of part-tiles which would be needed around the edges of the floor).

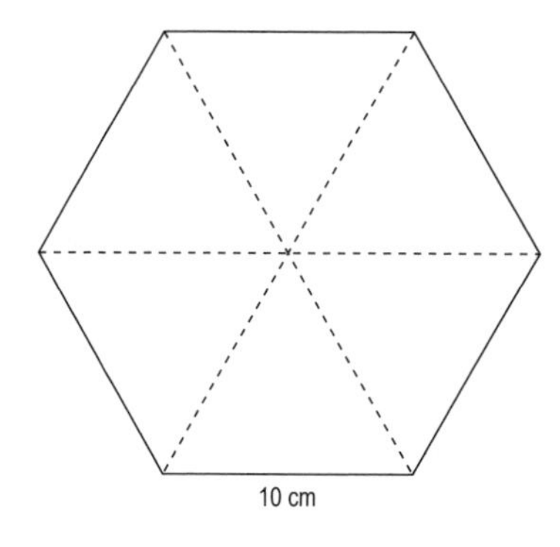

Fig 6.11

Notes – A regular hexagon is made up of 6 equilateral triangles; every angle in each triangle is 60°.

Find the area of each triangle, and hence find the area of the hexagon. Be careful with units:

1 m = 100 cm and 1 m^2 = 10000 cm^2.

Practical Assignment (6)

You have been asked to make a kite as a present for your friend's children.

The suggested kite is in the shape of two congruent triangles as shown in Fig 6.12.

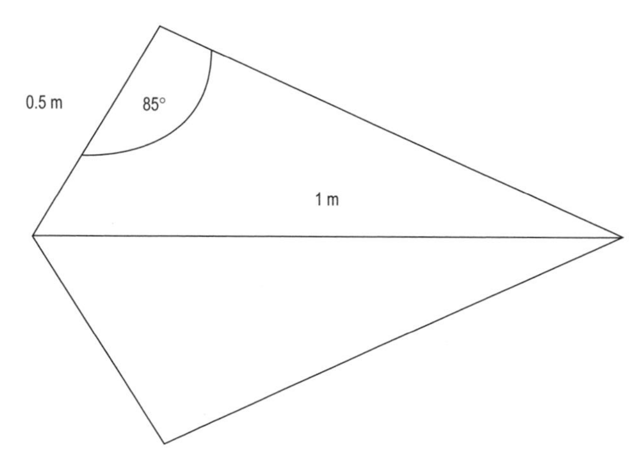

Fig 6.12

The fabric is sold by the square metre.

Find the area of the kite as shown, and investigate whether you could save any fabric by making changes in the dimensions or the angles, without changing the basic shape very much. The length of the kite, 1 metre, must be kept fixed.

Seminar discussion

Discuss how triangle measurement might be a useful skill in the design of kitchen equipment.

Study tip

Close your books and write down (i) the sine rule (ii) the cosine rule (iii) the formula for area of a triangle.

Check, and do this again until you can get them all right from memory.

7 Getting your bearings

One-minute overview

In this chapter we look at the use of trigonometry in solving problems relating to Bearings. This is a method of defining directions relative to North. Bearings problems provide further opportunities to practise using the sine rule and the cosine rule, often involving obtuse angles.

What is a bearing?

A bearing is a direction. Bearings are measured in degrees relative to North; the bearing due North is 0°.

We then turn clockwise from North, so due East is a bearing of 90°, due South is a bearing of 180°, and due West is a bearing of 270°.

Bearings are always written in 3 figures, so we write these bearings as 000°, 090°, 180° and 270°.

How to translate the wording into a diagram

In bearings problems, the position of a town, or a ship at sea, or some other object, is given like this:

"The bearing of A from B is 075°".

The important first step is to translate this statement into a correct diagram. After this stage, applying the trigonometry, along with some basic rules of geometry, is just the same as in the problems we have seen in earlier Chapters. So first we will practise drawing some diagrams corresponding to statements giving bearings. Note that we do not attempt to draw any diagram to scale. Angles are not measured, but just shown approximately correct, and there is no scale for the distances.

Examples (7.1)

1. "The bearing of A from B is 075°" means that A is on a bearing of 75° relative to B. So B is a fixed point of reference, and this is the first point to draw. Then, to show the bearing of A, we draw a line due North through B, turn clockwise through an angle of 75°, and draw a line from B in this direction. This is shown in Fig 7.1.

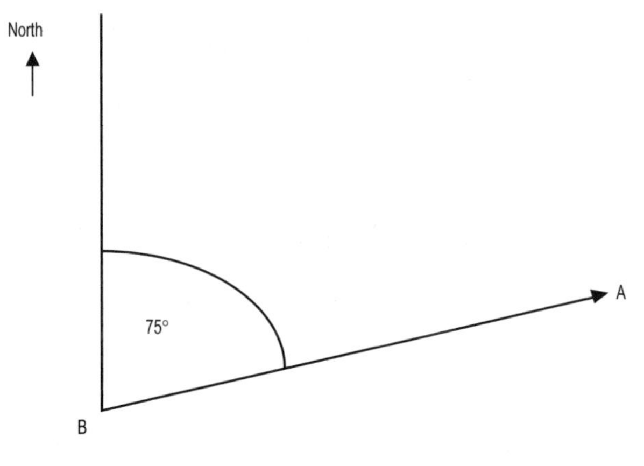

Fig 7.1

Note that we only know the bearing of A, i.e., the direction we need to take from the point B, to reach A.

The diagram shows a line from B pointing in the direction of A. We have not been told the distance from B to A in this example.

2. "B is 100 km from C on a bearing of 210°".

This time we are given the distance between the two points, as well as the bearing.

C is the fixed point of reference.

Fig 7.2 shows the positions of C and B.

The dotted line from C points due South.

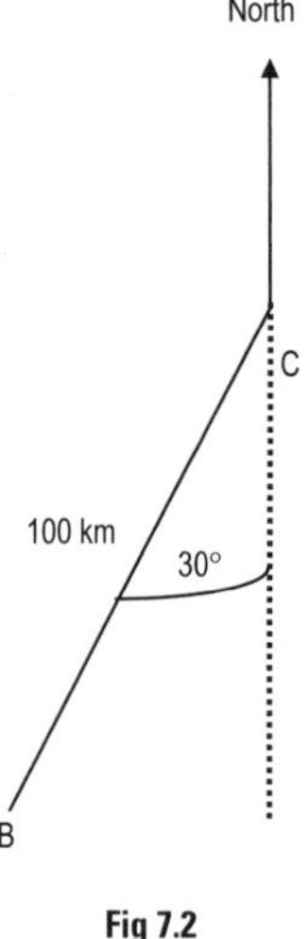

Fig 7.2

The angle between the vertical line through C, and the line CB, is 210° – 180° = 30°.

Practice Questions (7.1)

1. Draw a diagram for each of these situations.

   First decide which is the fixed point of reference.

   Draw and label this point, then draw a line due North through this point.

   Lastly, draw a line showing the bearing (and the distance, if given) of the second point.

   Label the relevant angle.

   (i) B is on a bearing of 68° from A.
   (ii) C is 64 km from A, on a bearing of 125°.
   (iii) E is 20 km from F, on a bearing of 260°.
   (iv) X is on bearing of 330° from W.

Note – diagrams for question 1 are not given in the Answers section, but questions (i) to (iii) are used in the examples below.

2. Convert these directions to bearings, and then draw the diagram:

   (i) H is North East of G
   (ii) P is South West of Q
   (iii) R is North West of S
   (iv) T is South East of U

Examples (7.2)

Now look at question 1(i) again. Suppose we are told that the shortest distance from B to a point C, which is due North of A, is 50 km.

The complete diagram is shown in Fig 7.3.

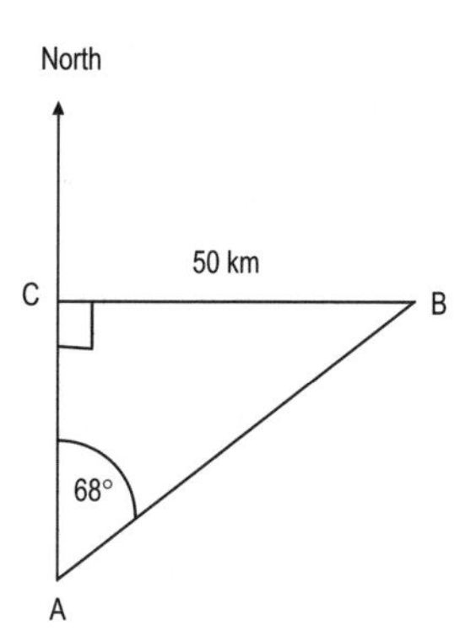

Fig 7.3

By applying simple trigonometry in triangle ABC, we can find the distance of B from A:

$$\sin 68° = 50/AB$$

$$AB = 50/\sin 68°$$

$$AB = 53.93 \text{ km (to 2 d.p.)}$$

And, if required, we would also be able to find the distance AC.

More often, we need to use the sine rule or the cosine rule, because bearings problems are not usually restricted to right-angled triangles. Look again at Question 1 (ii) in Practice Questions (7.1).

We will extend this question as follows:

Two ships set out from a point A. Ship B sails 80 km due South. Ship C sails 64 km on a bearing of 125° from A. What is the distance between the two ships?

Fig 7.4 shows the positions of A, B and C, with the known angles labelled:

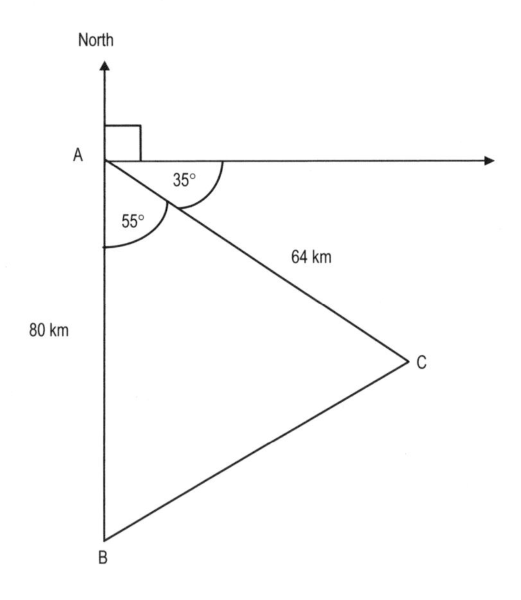

Fig 7.4

Note that the bearing 125° of ship C is split into two angles, 90° between the directions North and East, and 35° between the direction East and the line AC.

Subtracting 35° from 90° (the angle between due East and due South) gives $\angle BAC = 55°$.

This is the angle which enables us to find the distance BC; we now concentrate our attention on triangle ABC and apply the cosine rule.

Using the normal labelling convention, we have:

$$a^2 = b^2 + c^2 - 2bc \cos A$$

$$a^2 = 64^2 + 80^2 - 2(64)(80)\cos 55°$$

$$a^2 = 4622.61,\ a = 67.99 \text{ km}$$

So the distance between the two ships is 67.99 km.

Parallel lines and Alternate Angles

In some bearings problems we need to use the properties of parallel lines to work out angles.

The most useful rule is that of "Alternate Angles" (sometimes called Z angles).

Fig 7.5 shows the two equal angles α, which are created when a third line cuts a pair of parallel lines.

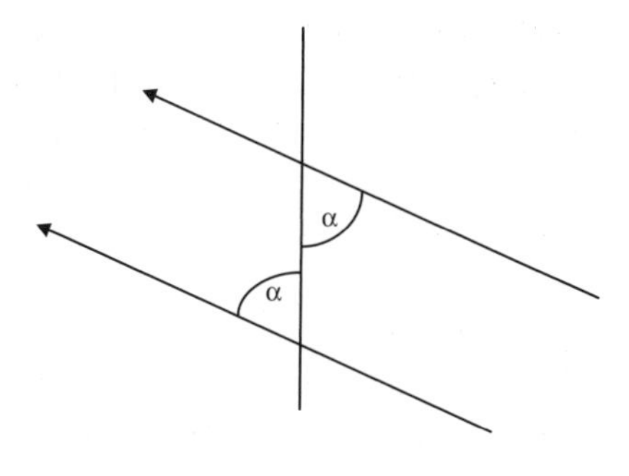

Fig 7.5

We use the property of Alternate Angles in the next example, which is an extension of question 1 (iii) from Practice Questions (7.1).

Example (7.3)

Sam cycles 20 km from his home in Forthampton to Edgefield, on a bearing of 260°. He then cycles 30 km to Greentown, which is on a bearing of 120° from Edgefield. He then cycles home. Sam's average cycling speed is 18 kph.

(i) Draw a diagram showing the positions of the three towns.
(ii) Calculate ∠FEG.
(iii) Use the cosine rule to find the distance from Forthampton to Greentown.
(iv) Use the sine rule to find ∠EFG and state the bearing of Greentown from Forthampton.
(v) Calculate Sam's total journey time in hours and minutes, to the nearest minute.

Fig 7.6 shows the completed diagram.

(i) To draw the diagram, first mark the position of F (Forthampton).

Then draw a line due North through F and the line from F to E (Edgefield).

The bearing of E from F is 260°. Measuring this angle clockwise from North takes us round to due South (180°) and leaves 80°, which is the angle between the line due South from F, and the line EF.

Mark the distance EF as 20 km. Next, we want to show the position of G (Greentown). The bearing 120° is equivalent to a turn of 30° in a clockwise direction from the line, which goes due East through E.

Mark the distance EG as 30 km.

(ii) If we can calculate ∠FEG, we can then use the cosine rule to find the distance FG.

We have two parallel lines, through the points E and F. Using the property of alternate angles, the angle between EF and the line due North through E, is 80°. The unlabelled portion of ∠FEG is therefore 10°, and the whole ∠FEG is 40°.

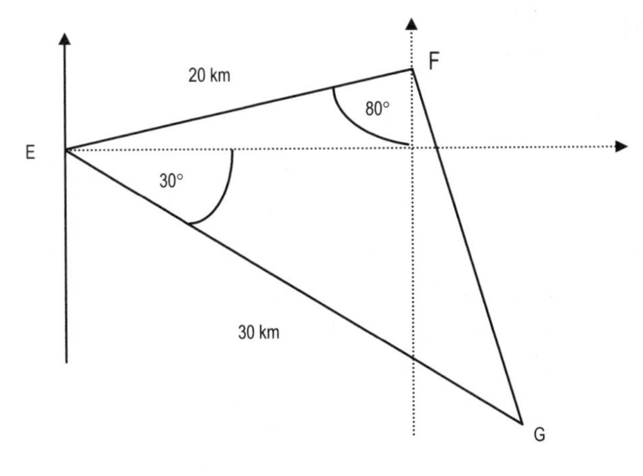

Fig 7.6

(iii) Now we can apply the cosine rule to find the distance FG:

$$FG^2 = 20^2 + 30^2 - 2(20)(30)\cos 40°$$

which gives $\quad FG = 19.51$ km.

(iv) To find the bearing of G from F, we use the sine rule to calculate ∠EFG:

$$\frac{\sin 40°}{19.51} = \frac{\sin EFG}{30}$$

Rearranging, $\sin EFG = \dfrac{30 \times \sin 40°}{19.51}$

$$\angle EFG = 81.3°$$

So the unlabelled portion of ∠EFG is 1.3°, and the bearing of G from F is (180° – 1.3°) = 178.7°

(v) The total distance travelled is 20 + 30 + 19.51 = 69.51 km.
Using the rule Time = Distance ÷ Speed, we have:
Time = 69.51/18 = 3.86 hours.
We need to convert 0.86 hours to minutes: 0.86 × 60 = 52 minutes to the nearest minute.
So Sam's journey time is 3 hours 52 minutes.

Tutorial

Progress Questions (7)

In all these questions, draw the diagram, and label all the angles and distances, before you start the calculations.

1. A plane flies 500 km from A to B on a bearing of 140°. It then flies 400 km from B to C on a bearing of 020°.
   - (i) What is the bearing of A from B?
   - (ii) What is the bearing of B from C?
   - (iii) Calculate the distance from C to A.
   - (iv) Find the bearing of A from C.
2. A group of walkers set off from their hostel H and walk 15 km on a bearing of 021° to point L where they stop for lunch. After lunch, they set off again, on a bearing of 290° and walk 10.5 km to point B where they catch a bus back to the hostel.
   - (i) Calculate the length of the bus journey.
   - (ii) Find the bearing of the hostel from the bus stop.
3. A ship S leaves port P on a bearing of 130°. When it has travelled 50 km, a member of the crew is taken seriously ill and the captain drops anchor and radios for a helicopter ambulance, which sets off from its departure point H, 40 km North East of P. The helicopter flies at 200 kph. Calculate, to the nearest minute, the time the helicopter takes to reach the ship.
4. John walks 3 km South West from his house H to his office O. The journey takes 45 minutes.

   John plans to do this on every workday. But if he gets up late, he takes the bus from the bus stop outside his house to point B, which is 0.4 km from his office on a bearing of 100°. The bus travels at 21 kph. John then walks at his normal speed from point B to the office. Assuming that John has to wait a negligible amount of time for the bus to arrive, how much extra time, to the nearest minute, can he spend in bed if he is feeling too tired to walk to work?

Practical Assignment (7)

Using an ordinary compass, choose a point in your room where you can walk forwards on a bearing of, say, 050°. Turn round and read the bearing which you need to take to return to your starting point.

Check that the difference between your two bearings is 180°.

Seminar discussion

Are bearings a more reliable method for giving directions than compass directions such as NNW and South 40° E?

Study tip

There are no new formulae or rules to learn in this Chapter, so your next best step is to prepare for Chapter 8 by doing some revision from Chapter 5.

8 Making different waves

One-minute overview

In this chapter we explore the properties of the three trig graphs. We then look at transformations of these graphs (translations, reflections, and one-way stretches) including some examples involving more than one transformation. We use these graphs to find the maximum and minimum values of trig functions, and to read off values at particular points.

The sin graph again

From your work in Chapter 5, you have seen that the graph of the sin function in the range 0° to 360° looks like the curve shown in Fig 8.1:

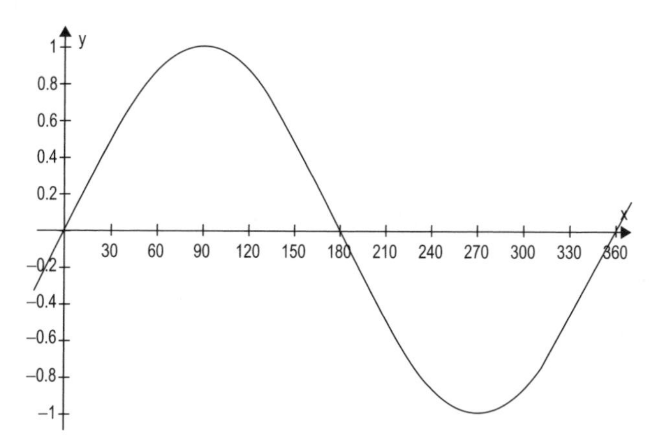

Fig 8.1

There are some important points to note from this graph:

1. The maximum value of $\sin x$ is 1, and this occurs at 90° (or $\pi/2$ radians).

2. The minimum value of sin x is –1, and this occurs at 270° (or $3\pi/2$ radians).
3. There are three points at which sin x = 0, and these are 0°, 180°, and 360°.

We saw in Chapter 5 that any angle outside the range 0° to 360° is equivalent to another angle inside this range. For example, 370° is equivalent to 10°, 380° is equivalent to 20°, –10° is equivalent to 350°, –20° is equivalent to 340°, etc. Therefore, the sins of these equivalent angles are equal: sin 370° = sin 10°, etc.

The shape of the sin graph outside the range 360° to 720°, is the same as it is inside this range. Because of this property, we describe the sin graph as periodic, or cyclic. A graph of the sin function over a wider range, shown in Fig 8.2, explains the use of the term "sin wave":

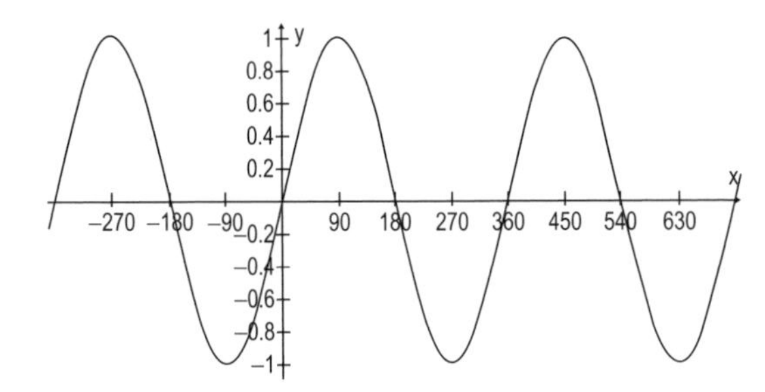

Fig 8.2

Practice Questions (8.1)

1. Sketch the graph of y = cos x in the range 0° to 360°.

   Giving answers in both degrees and radians, write down the angles for which:

   (i) cos x = 1
   (ii) cos x = –1
   (iii) cos x = 0
2. Sketch the graph of y = cos x in the range –360° to 720°.
3. Sketch the graph of y = tan x in the range 0° to 360°.

   Giving answers in both degrees and radians, write down the angles for which:

   (i) tan x = 0
   (ii) tan x is undefined

4. Sketch the graph of $y = \tan x$ in the range $-360°$ to $720°$.
5. The period of a cyclic or periodic function is the width of the range of values of x over which the repeating shape occurs once. For the function $\sin x$, the period is $360°$.

   What is the period of the function (i) $\cos x$ (ii) $\tan x$?

Transformations of trig graphs

(i) Translations

Let's look at the function $y = \sin x + 1$.

To sketch the graph of this function, we just add the number 1 to each value of $\sin x$.

So for $x = 0°$, $y = (\sin 0°) + 1 = 0 + 1 = 1$, so this gives us the point $(0°, 1)$.

And for $x = 90°$, $y = (\sin 90°) + 1 = 1 + 1 = 2$, so this gives us the point $(90°, 2)$.

We see that the sin curve simply shifts upwards by 1 unit, giving us the graph shown in Fig 8.3:

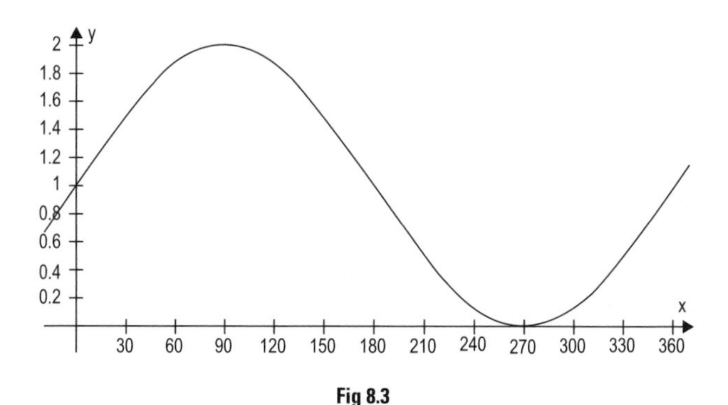

Fig 8.3

Note from this graph that the maximum and minimum values of $y = \sin x + 1$ are 2 and 0.

In general, the function $y = \sin x + a$ where a is any constant, has the same shape as the function $y = \sin x$, and represents a vertical translation, up or down a units, according to whether a is positive or negative, with the maximum and minimum values of the function adjusted accordingly.

A horizontal translation of the graph of $y = \sin x$ is given by any function in the form $y = \sin(x \pm \alpha)$.

Let's look at the graph of $y = \sin(x + 45°)$.

Here are some points which will be on the graph:

$$x = 0°,\ y = \sin(0° + 45°) = \sin 45° = 0.707$$

$$x = 45°,\ y = \sin(45° + 45°) = \sin 90° = 1$$

$$x = 90°,\ y = \sin(90° + 45°) = \sin 135° = 0.707$$

Continuing in this way enables us to see that the graph of $y = \sin(x + 45°)$ is as shown in Fig 8.4:

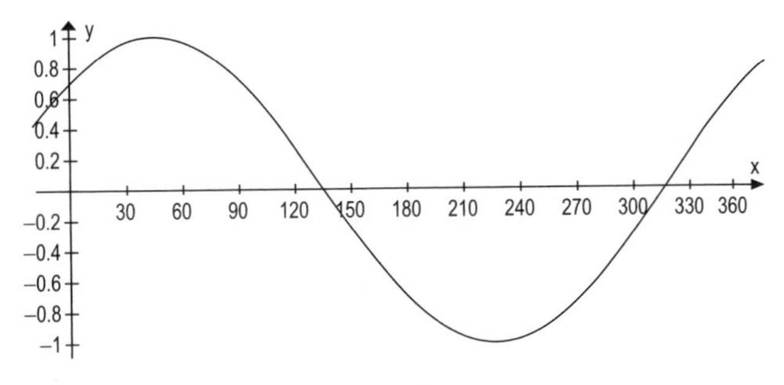

Fig 8.4

And comparing this graph with the graph of $y = \sin x$, we see that the graph of $y = \sin(x + 45°)$ is a left shift of the graph of $y = \sin x$, by 45°. For example, the point (45°, 1) on the graph in Fig 8.4 corresponds to the point (90°, 1) on the graph of $y = \sin x$.

In general, when α is any positive angle:

The graph of the function $y = \sin(x + \alpha)$ represents a translation to the left, by α degress, and the graph of the function $y = \sin(x - \alpha)$, represents a translation to the right, by α degrees.

Caution:
Take care to distinguish between the function $y = \sin x + a$ and the function $y = \sin(x + \alpha)$. "$\sin x + a$" means "take the value of $\sin x$ and add the constant value a".

This type of function is also often written $y = a + \sin x$.

But "$\sin (x + \alpha)$" means "add the angles α and x and then find the sin of the resulting angle".

Practice Questions (8.2)

1. Sketch the graphs of (i) $y = \sin x - 2$ (ii) $y = 2 + \sin x$.
   For each of these graphs, write down the maximum and minimum values of y.
2. Sketch the graphs of (i) $y = \sin (x + 90°)$
   (ii) $y = \sin (x - 60°)$ (iii) $\sin (x + 30°)$
3. Sketch the graphs of (i) $y = \cos x - 1$ (ii) $y = 1 + \cos x$.
   For each of these graphs, write down the maximum and minimum values of y.

(ii) Reflections

Let's look at the function $y = -\sin x$. The values of this function have opposite signs to the values of the function $y = \sin x$. For example, $\sin 90° = 1$, and $-\sin 90° = -1$. The graph of $y = -\sin x$ is shown in Fig 8.5:

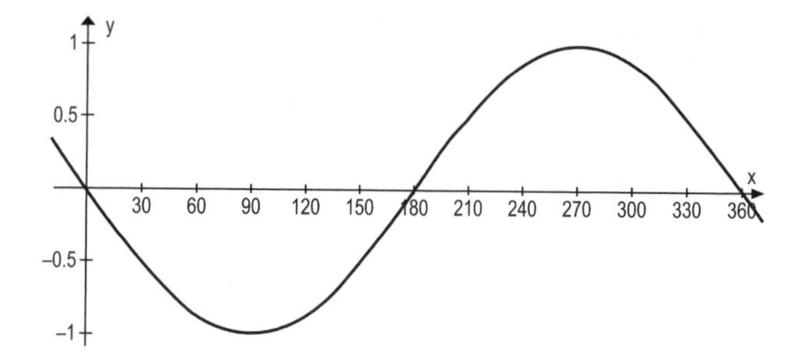

Fig 8.5

We see that reversing the signs has the effect of a reflection in the x-axis.

Now let's look at the function $y = \sin (-x)$:

When $x = 0°$, $y = \sin (-0°) = \sin 0° = 0$.
When $x = 90°$, $y = \sin (-90°) = \sin 270° = -1$.
When $x = 180°$, $y = \sin (-180°) = \sin 180° = 0$.
When $x = 270°$, $y = \sin (-270°) = \sin 90° = 1$.

So we have the points (0, 0), (90°, –1), (180°, 0), (270°, 1) and we can see that the graph of $y = \sin(-x)$ is in fact the same as graph of $y = -\sin x$, i.e. a reflection of the graph of $y = \sin x$ in the x-axis. This can also be seen as a reflection in the y-axis.

BUT note that the effect of a reflection in the x-axis and a reflection in the y-axis *happen to be the same* in the case of the function $y = \sin x$, and these will not be the same, in general, for other functions. To avoid problems later, we will specify these reflections as:

$y = -\sin x$	Reflection of $y = \sin x$ in the x-axis.
$y = \sin(-x)$	Reflection of $y = \sin x$ in the y-axis.

(iii) One-way stretches

Let's look at the function $y = 2 \sin x$. To sketch the graph of this function, we need to multiply all the values of $\sin x$ by 2. The result is a sin curve which is stretched by a factor of 2 parallel to the y-axis.

The maximum and minimum values of this function are 2 and –2. The graph is shown in Fig 8.6.

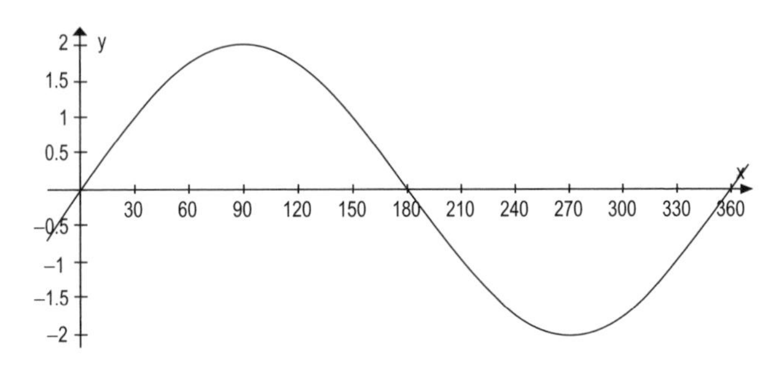

Fig 8.6

Now let's look at the function $y = \sin(2x)$.

The x-values 0, 45°, 90°, 135° and 180° give us the points (0, 0), (45°, 1), (90°, 0), (135°, –1), (180°, 0).

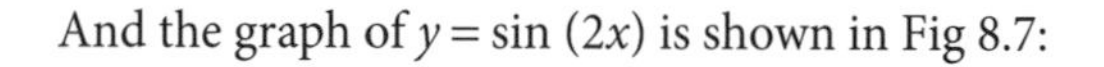

And the graph of $y = \sin(2x)$ is shown in Fig 8.7:

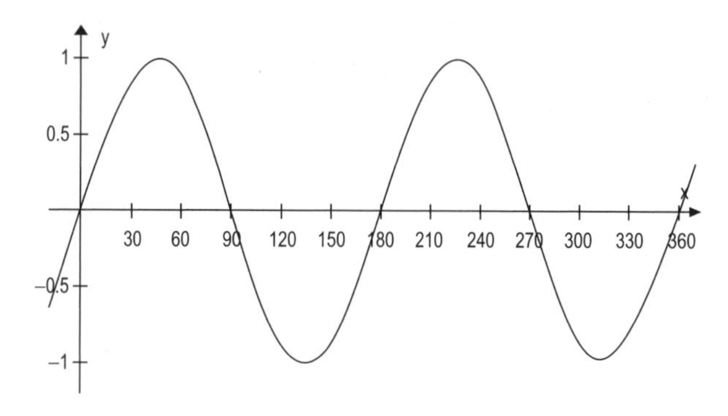

Fig 8.7

The graph of $y = \sin 2x$ does not *appear* to be a stretch of the graph of $y = \sin x$. It is in fact a sin wave with period 180°, i.e. the frequency of the wave is doubled. Mathematically, we describe this as a stretch with factor ½.

We would therefore expect that the graph of $y = \sin \frac{1}{2} x$ will be a stretch of $y = \sin x$, with factor 2.

Summary of transformations

This table summarises the transformations we have covered in this chapter.

You should find this table useful with the questions that follow it.

Vertical translation, k units	Horizontal translation, k degrees	Reflection in *x*-axis	Reflection in *y*-axis	Vertical stretch, factor k	Horizontal stretch, factor 1/k
$y = k + \sin x$	$y = \sin(x - k)$	$y = -\sin x$	$y = \sin(-x)$	$y = k \sin x$	$y = \sin kx$

Practice Questions (8.3)

For each of the following functions, first describe the transformation, then sketch the graph.

1. $y = 3 + \cos x$
2. $y = \cos (x - 90°)$
3. $y = 4 \cos x$
4. $y = \cos 2x$
5. $y = 1 + \tan x$
6. $y = \tan (x + 45°)$
7. $y = \tan (x/3)$
8. $y = \sin x - 2$ (i.e. $y = -2 + \sin x$).
9. Sketch the graph of the function $y = 2 \cos x - 1$, which can be obtained from the graph of $y = \cos x$ by a vertical stretch, factor 2, followed by a vertical translation, down 1 unit.
10. Sketch the graph of the function $y = \tan (\frac{1}{2}x - 90°)$, which can be obtained from the graph of $y = \tan x$ by a horizontal stretch, factor ½, followed by a horizontal translation, 90° to the right.
11. Write down the transformations, in the correct order, needed to obtain the graphs of these functions:
    (i) $y = 2 + \cos 3x$
    (ii) $y = \frac{1}{2} \tan (x - 30°)$
    (iii) $y = \sin (2x - 60°)$.

Tutorial

Progress Questions (8)

1. Sketch the graph of $y = \sin x$ in the range $-360°$ to $360°$. Use your graph to write down <u>four</u> angles for which
   (i) $\sin x = 0.5$
   (ii) $\sin x = -0.5$
2. Sketch the graph of $y = \tan x$ in the range $-180°$ to $180°$. Use your graph to write down <u>two</u> angles for which
   (i) $\tan x = 1$
   (ii) $\tan x = -1$
3. Use a graphical calculator to display the function $y = \sin(x + 60°)$ in the range $0°$ to $360°$.
   Read off the values for which $\sin(x + 60°) = 0.22$
4. Use a graphical calculator to display the function $y = \cos(x - 24°)$ in the range $0°$ to $360°$.
   Read off the values for which $\cos(x - 24°) = -0.5$
5. Write down the transformations needed to obtain the graph of the function $y = 1 - \sin x$.
   Then sketch the graph and write down the values of x for which:
   (i) $1 - \sin x = 2$
   (ii) $1 - \sin x = 0$
   (iii) $1 - \sin x = 0.5$
6. Sketch the function $y = \cos(-x)$ and comment on what you notice.
7. Write down the maximum and minimum values of
   (i) $2 + \cos x$
   (ii) $1 - \sin x$
   (iii) $\frac{1}{4} \sin x$

Practical Assignment (8)

You will need to sketch trig curves many more times in your study of this subject.

Find a way to reproduce these curves to save yourself the time it takes to keep drawing them.

Seminar discussion

Many people find aspects of mathematics aesthetically pleasing, and the trigonometric curves have inspired designers in a variety of ways. Suggest some appropriate ways in which these curves could be used to decorate everyday household items.

Study tip

Look again at the summary table of transformations, close the book and write the table down from memory.

9 How to solve trig equations

One-minute overview

In this chapter we learn the method for solving trig equations which involve functions of the type whose graphs we studied in Chapter 8. We use our knowledge of the properties of trig functions to develop a method for finding the general solution of a trig equation. The general solution is the solution that includes all the possible angles which satisfy the equation.

What is a trig equation?

A trig equation equates a trig expression, containing one or more trig functions, with a constant value.

The solution of the equation is a set of angles, all of which satisfy the equation.

You have solved some simple trig equations by graphical methods in the Progress Questions in Chapter 8.

We now look more closely at methods of solving trig equations. You will need to use the inverse trig functions on your calculator. A standard scientific calculator gives one angle, so we need to use our knowledge of trig functions to find all the other angles in the required range. The one angle which the calculator provides is the angle closest to 0°; in some cases this angle is negative.

Common mathematical letters

Different letters are commonly used to represent angles. We have used the letter α (alpha) and we have used the letters A, B and C and others when working with triangles. The letter x is often used to represent the angle when working with trig graphs, as we did in Chapter 8. The letter θ (theta) is also commonly used in trigonometry, particularly in trig equations.

Example (9.1)

Solve the trig equation: $\cos \theta = 0.4$ giving solutions in the range 0° to 360°.

We find the first solution to this equation from the inverse cos function on the calculator, $\theta = 66.4°$.

Sketching the graph of $y = \cos \theta$ in the range 0° to 360° will help us to find the second solution:

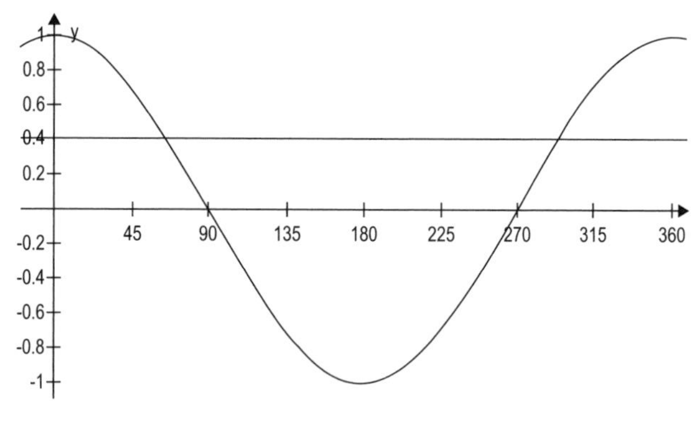

Fig 9.1

In Fig 9.1, the line $y = 0.4$ cuts the curve $y = \cos \theta$ at two points.

The first is where $\theta = 66.4°$ (our calculator value) and the second is where $\theta = 360 - 66.4 = 293.6°$.

If you are completely confident that you can apply the rules you learnt in Chapter 5, then you may not need to sketch a graph every time you are solving an equation, but if in any doubt, a graph will always help.

The solutions we have found to the equation $\cos \theta = 0.4$ are the only two solutions in the range 0° to 360°.

If we are asked to find the general solution, we have to find all the angles for which $\cos \theta = 0.4$.

There are two such angles in every period of the function, i.e. two angles in the range −360° to 0°, two angles in the range 360° to 720°, etc. We can find all these angles by adding or subtracting 360° to or from the two solutions we already have: –293.6°, –66.4°, 426.4°, 653.6°, etc. However, this will give us an infinite list of angles. So we replace this numerical list of angles with a formula which will neatly express the entire set of solutions:

$$\theta = 66.4° + (360° \times n) \quad \text{where } n = \pm 1, \pm 2, \pm 3\ldots$$

$$\theta = 293.6° + (360° \times n) \quad \text{where } n = \pm 1, \pm 2, \pm 3\ldots$$

Example (9.2)

Solve the trig equation: $\sin 2\theta = 0.5$ giving solutions in the range 0° to 360° and the general solution.

The calculator gives inverse sin (0.5) = 30°. So we write $2\theta = 30°$, then $\theta = 15°$.

Sketching the graph of $y = \sin 2\theta$ will help us to find other solutions in the range 0° to 360°:

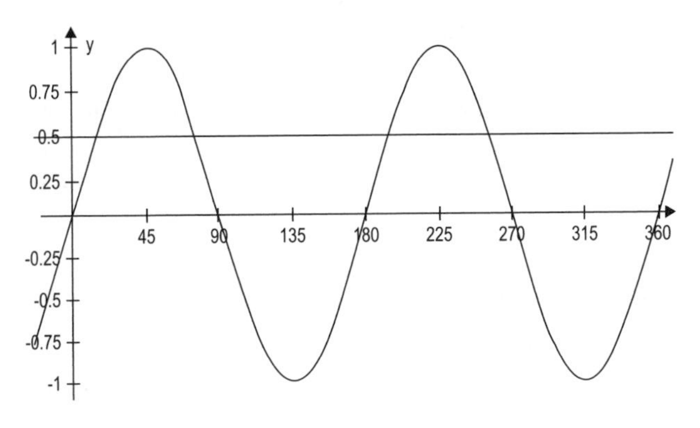

Fig 9.2

The line $y = 0.5$ cuts the curve $y = \sin 2\theta$ in four places, at $\theta = 15°$, (90° – 15°), (180° + 15°), (270° – 15°).

So the solutions in the range 0° to 360° are θ = 15°, 75°, 195°, 255°.

The period of the curve $y = \sin 2\theta$ is 180° and the set of solutions repeats at this interval.

So the general solution is $\theta = 15° \pm (180° \times n)$, $75° \pm (180° \times n)$.

It is important to understand that the number of solutions in the range 0° to 360° depends on the period of the function. Because the function $y = \sin 2\theta$ has period 180° (i.e. the wave occurs twice in the range 0° to 360°) there are four solutions in the range 0° to 360°. Therefore, we would expect six solutions to an equation such as $\sin 3\theta = k$, but only one solution to an equation such as $\sin \frac{1}{2}\theta = k$.

In fact, since the curve $y = \frac{1}{2} \sin \theta$ has only the positive half of the wave in the range 0° to 360°, the equation $\sin \frac{1}{2}\theta = k$ will have one solution in the range 0° to 360° if k is positive, but no solution in this range if k is negative. This is relevant in question (iii) below.

Practice Questions (9.1)

Do not overlook the fact that answers to an equation can always be checked by putting your answer back into the equation. When you have found a value of θ for question (i), check that $\tan \theta = 1.5$. If not, check your work and look for the error. The Answers section is provided for completeness, but you should get used to checking your answers first.

Solve these equations in the range 0° to 360°, and then write down the general solutions:

(i) $\tan \theta = 1.5$	(ii) $\cos 3\theta = 0.8$
(iii) $\sin \frac{1}{2}\theta = -0.35$	(iv) $\tan 2\theta = -1$

Examples (9.3)

Here are some examples where we solve equations without using graphs, including some cases requiring solutions in the range −180° to 180°:

(i) Solve the trig equation $\cos(2\theta + 30°) = 0.6$ giving all the solutions in the range 0° to 360°.

Using the inverse cos function on the calculator gives $2\theta + 30° = 53.1°$ (to 1 d.p.).

The second angle in the range 0° to 360° is $(360° - 53.1°) = 306.9°$.

So now we have two equations:

$$2\theta + 30° = 53.1° \quad \text{and} \quad 2\theta + 30° = 306.9°$$

which simplify to:

$$2\theta = 23.1° \quad \text{and} \quad 2\theta = 276.9°$$

so we have two solutions:

$$\theta = 11.6^\circ \quad \text{and} \quad \theta = 138.4^\circ$$

(working to 1. d.p).

The period of the function $y = \cos(2\theta + 30^\circ)$ is 180°, so there are two more solutions in the range 0° to 360°, and these are $(11.6^\circ + 180^\circ) = 191.6^\circ$ and $(138.4^\circ + 180^\circ) = 318.4^\circ$.

The full set of solutions is therefore: $\theta = 11.6^\circ, 138.4^\circ, 191.6^\circ, 318.4^\circ$.

(ii) Solve the trig equation $\tan(\theta - 60^\circ) = -2$ giving all solutions in the range -180° to 180°:

The inverse tan function gives: $(\theta - 60^\circ) = -63.4^\circ$.

Hence $\theta = -3.4^\circ$

The function $y = \tan x$ has period 180°, so the solutions in the range -180° to 180° are:

$$\theta = -3.4^\circ \quad \text{and} \quad \theta = -3.4^\circ + 180^\circ = 176.6^\circ.$$

(iii) Solve the trig equation $2\sin(\theta - 80^\circ) = -0.1$ giving all the solutions in the range -180° to 180°.

In this example, we first need to divide both sides of the equation by 2:

$$\sin(\theta - 80^\circ) = -0.05$$

Now we can use the inverse sin function, and we get:

$$\theta - 80^\circ = -2.9^\circ \quad \text{and} \quad \theta - 80^\circ = (180^\circ - (-2.9^\circ)).$$

So our two solutions are $\quad \theta = 77.1^\circ \quad$ and $\quad \theta = 262.9^\circ$

Practice Questions (9.2)

1. Solve these equations giving all solutions in the range 0° to 360°:

   (i) $\sin(\frac{1}{2}\theta - 20^\circ) = 0.7$ (ii) $\tan(3\theta + 70^\circ) = 2.4$

   (iii) $\cos(\frac{1}{4}\theta + 30^\circ) = 0.4$

2. Solve these equations giving all solutions in the range 0° to 360°.

   Note that as in Example 9.3 (iii) above, you will first need to multiply or divide both sides by a constant before you can apply an inverse trig function:

   (i) $3\tan\theta = 3$ (ii) $\frac{1}{2}\cos 2\theta = 0.25$

   (iii) $4\sin(\theta - 30^\circ) = 1$

Working with radians

Sometimes we are required to solve a trig equation giving the solution in radians.

The basic process is just the same, but this table of equivalent values of radians and degrees will be useful:

	1st quadrant	2nd quadrant	3rd quadrant	4th quadrant
Degrees	0° to 90°	90° to 180°	180° to 270°	270° to 360°
Radians in terms of π	0 to $\pi/2$	$\pi/2$ to π	π to $3\pi/2$	$3\pi/2$ to 2π
Radians in decimals	0. to 1.571	1.571 to 3.142	3.142 to 4.712	4.712 to 6.284

Example (9.4)

Solve the trig equation $\cos 2\theta = 0.76$, giving all solutions in the range 0 to 2π, and the general solution.

We set the calculator to work in radians, and the inverse cos function gives

$$2\theta = 0.707 \text{ (to 3 d.p.)}$$

The cos function is positive in the fourth quadrant, so the second solution is:

$$2\theta = (2\pi - 0.707) = 5.576$$

Dividing by 2, $\quad \theta = 0.354 \quad$ and $\quad 2.788$

Since the function $\cos 2\theta$ has period π radians, there will be two more solutions in the range 0 to 2π.

These are $\theta = (0.354 + \pi)$ and $\theta = (2.788 + \pi)$, i.e. $\theta = 3.500$ and 5.930.

The full set of solutions in the range 0 to 2π is $\theta = 0.354$, 2.788, 3.500, 5.930.

The general solution is $\theta = 0.354 + n\pi$, $\theta = 2.788 + n\pi$.

Practice Questions (9.3)

Solve these equations, giving solutions in the range $-\pi$ to π and the general solution:

(i) $\sin \frac{1}{2}\theta = 0$ (ii) $4 \tan 2\theta = 5$

(iii) $3 \cos (\theta + \pi/2) = -1.2$

The functions $\sin^2\theta$, $\cos^2\theta$ and $\tan^2\theta$

Here is another point where we have to be careful to understand conventional notation. "$\sin^2\theta$" means the same as $(\sin\theta)^2$ and similarly, $\cos^2\theta$ means $(\cos\theta)^2$, and $\tan^2\theta$ means $(\tan\theta)^2$. It is normal to write $\sin^2\theta$ in place of $(\sin\theta)^2$. In the next chapter we will be working with squared trig functions, but for now we will solve some trig equations using this notation.

Examples (9.5)

(i) Solve the equation $\cos^2\theta = 0.25$ giving solutions in the range 0 to π.

The first step is to take the square root of each side, giving $\cos\theta = \pm 0.5$.

So the equation $\cos^2\theta = 0.25$ is actually two equations: $\cos\theta = 0.5$ and $\cos\theta = -0.5$.

We solve the two equations one at a time:

$\cos\theta = 0.5$: $\theta = 1.047$ is the only solution in the range 0 to π.

$\cos\theta = -0.5$: $\theta = 2.094$ is the only solution in the range 0 to π.

The two solutions to the equation $\cos^2\theta = 0.25$ are $\theta = 1.047$, $\theta = 2.094$.

(ii) Solve the equation $2\sin^2\theta - 1 = 0$ giving solutions in the range 0 to 2π.

First we need to rearrange the equation: $\sin^2\theta = 0.5$

Next step is to take the square root: $\sin\theta = \pm 0.707$

Now we have two equations to solve: $\sin\theta = 0.707$ and $\sin\theta = -0.707$

The first equation has solutions $\theta = 0.785$ and $\theta = 2.356$

The second equation has solutions $\theta = 3.927$ and $\theta = 5.498$

So the solutions to the equation $2\sin^2\theta - 1 = 0$ in the range 0 to 2π are $\theta = 0.785, 2.356, 3.927, 5.498$.

Tutorial

Progress Questions (9)

1. For each of these equations, give a clear mathematical reason why there is no solution:
   (i) $\sin\theta = 2$ (ii) $\tan^2\theta = -2$ (iii) $\frac{1}{4}\cos\theta = 0.3$
2. For each of these equations, explain why there is no solution in the given range:
   (i) $\cos\frac{1}{2}\theta = -0.2$, range $-\pi$ to π
   (ii) $\sin(-\frac{1}{2}\theta) = 0.5$, range 0 to 2π
3. Solve these equations giving all the solutions in the range 0° to 360°:
   (i) $\tan^2 2\theta = 1$ (ii) $\cos(\theta + 60°) = 1$
   (iii) $2\cos^2\theta - 1 = 0$
4. Solve these equations giving the general solution in radians:
   (i) $\tan(\theta - \pi) = 2$ (ii) $4\sin^2\theta = 0.09$
   (iii) $2\cos\frac{1}{2}\theta = -1$

Practical Assignment (9)

Use a graphical calculator to find, in degrees, two values of θ for which $\sin\theta = 0.45$.

Seminar discussion

Describe phenomena in the natural world where an understanding of periodic functions is important.

Study tip

By now you should be confident with sketching trig functions, understanding the effects of transformations, and applying all these skills to solving trig equations. If you are still feeling at all shaky, revise all of Chapters 5, 8 and 9.

10 Understanding identities

One-minute overview

In this chapter we first define the term "identity" and learn to distinguish an identity from an equation. We look at more relationships between trig functions, and we learn to manipulate trig expressions in order to prove identities. Finally we see how to use identities to simplify and solve trig equations.

What is an identity?

An identity is a mathematical statement which is always true irrespective of the values of the variables.

This is distinct from an equation, which is a mathematical statement which is true only if the variable(s) take specific values.

Examples (10.1)

Here is an algebraic equation: $3x + 2 = 11$.

This equation holds true only if $x = 3$. For any other value of x, the statement "$3x + 2 = 11$" is false.

Here is an algebraic identity: $3(x + 1) \equiv 3x + 3$.

These two expressions are always equal, whatever the value of x.

$3(x + 1)$ is just a different way of writing $3x + 3$.

Notice the use of this special equals sign $\equiv$ with its three lines rather than two.

The sign $\equiv$ means "is identically equal to", i.e. the two expressions are always equal.

Unfortunately, the usual equals sign $=$ is often used when the sign $\equiv$ should be used.

However, we will be strict here and use the sign $\equiv$ whenever it applies.

Here is a trig equation: $\cos\theta = 1$.

This equation only holds true if $\theta = (n \times 360°)$. This is the general solution to this equation.

Here is a trig identity: $\cos\theta \equiv \sin(90° - \theta)$.

This statement is true for any angle θ.

Practice Questions (10.1)

Complete these statements, inserting either = or ≡ in the box according to whether the statement is an equation or an identity:

1. $(x-1)(x+1)$ ☐ $x^2 - 1$
2. $x - 1$ ☐ 100
3. $\tan\theta$ ☐ 1
4. $\sin\theta$ ☐ $\sin(180° - \theta)$
5. $\cos\theta$ ☐ $\cos(-\theta)$
6. $\sin^2\theta$ ☐ 1

Two important trig identities

There are two trig identities which we can derive from the trig ratios in a right-angled triangle.

In Fig 10.1 we have:

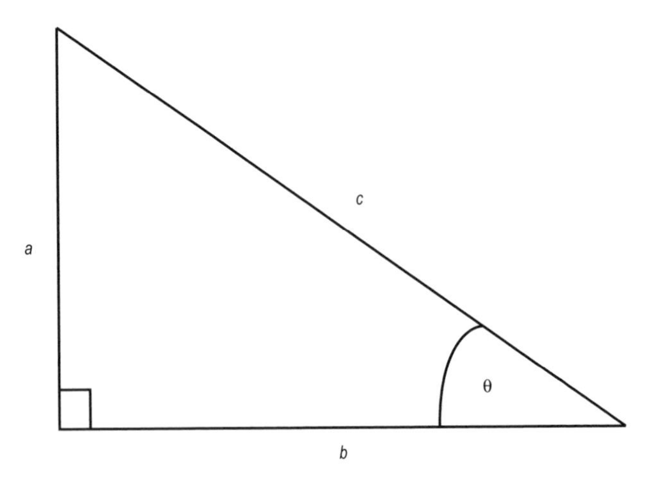

Fig 10.1

$$\sin\theta = \frac{a}{c}$$

$$\text{and} \quad \cos\theta = \frac{b}{c}$$

Dividing sin θ by cos θ, we have:

$$\frac{\sin\theta}{\cos\theta} = \frac{a}{c} \div \frac{b}{c}$$

which simplifies to:

$$\frac{\sin\theta}{\cos\theta} = \frac{a}{b}$$

Since $\tan\theta = \frac{a}{b}$ we see that

$$\tan\theta = \frac{\sin\theta}{\cos\theta}$$

and this is our first trig identity. Since this relationship is true for all values of θ, we use the sign ≡ :

$$\tan\theta \equiv \frac{\sin\theta}{\cos\theta} \quad \text{..............Identity 1}$$

We derive our next trig identity by applying Pythagoras to the triangle in Fig 10.1:

$$a^2 + b^2 = c^2$$

Dividing every term by c^2 gives:

$$\frac{a^2}{c^2} + \frac{b^2}{c^2} = 1$$

Replacing $\frac{a}{c}$ by sin θ and $\frac{b}{c}$ by cos θ gives:

$$\sin^2\theta + \cos^2\theta = 1$$

And again we use the sign ≡ to indicate that this relationship is true for all values of θ:

$$\sin^2\theta + \cos^2\theta \equiv 1 \quad \text{..........Identity 2}$$

These identities are very important, and you should memorise them now. You will need to use them to prove other relationships between trig functions, and to solve some types of trig equations.

The addition formulae

These are a group of formulae which are also regularly used in the proof of other identities and in the solving of equations. We don't have room in this book to write out the proofs, which are done by looking at the trig ratios in two overlapping triangles. If you would like to read the proofs, you could look at:

http://mathforum.org/library/drmath/view/54051.html
or any similar website.

Here are the addition formulae, in which A and B are any two angles:

1. $\sin(A + B) \equiv \sin A \cos B + \cos A \sin B$
2. $\sin(A - B) \equiv \sin A \cos B - \cos A \sin B$
3. $\cos(A + B) \equiv \cos A \cos B - \sin A \sin B$
4. $\cos(A - B) \equiv \cos A \cos B + \sin A \sin B$
5. $\tan(A + B) \equiv \dfrac{\tan A + \tan B}{1 - \tan A \tan B}$
6. $\tan(A - B) \equiv \dfrac{\tan A - \tan B}{1 + \tan A \tan B}$

Now we'll look at two numerical examples, which will help you understand how these formulae work. First note the use of the abbreviations LHS (left hand side) and RHS (right hand side) which are commonly used when working with identities.

Examples (10.2)

With A = 30° and B = 60° in addition formula 1, we have

$$\text{LHS} = \sin(30° + 60°) = \sin 90° = 1$$

$$\text{RHS} = \sin 30° \cos 60° + \cos 30° \sin 60°$$

$$= \frac{1}{2} \times \frac{1}{2} + \frac{\sqrt{3}}{2} \times \frac{\sqrt{3}}{2} = \frac{1}{4} + \frac{3}{4} = 1$$

With A = 45° and B = 45° in addition formula 3, we have

$$\text{LHS} = \cos(45° + 45°) = \cos 90° = 0$$

$$\text{RHS} = \cos 45° \times \cos 45° - \sin 45° \sin 45°$$

$$= \frac{1}{\sqrt{2}} \times \frac{1}{\sqrt{2}} - \frac{1}{\sqrt{2}} \times \frac{1}{\sqrt{2}} = \frac{1}{2} - \frac{1}{2} = 0$$

(If you have forgotten these sin and cos values, look back to "Three Special Angles" in Chapter 3).

Practice Questions (10.2)

1. Choose any angle θ and use your calculator to show that $\sin^2\theta + \cos^2\theta = 1$ for your value of θ.
2. Choose any angle θ and use your calculator to show that $\tan\theta = \frac{\sin\theta}{\cos\theta}$ for your value of θ.
3. Using the method shown in Examples 10.2, verify addition formula 2 with A = 90° and B = 30°.
   You should not need to use a calculator.
4. Which of the addition formulae is the same as Identity 2, when A = B?
5. Choose angles A and B so that A + B < 90°, and verify addition formula 5 for these two angles.

The double-angle formulae

Addition formulae 1, 3 and 5 are especially useful in the cases where A and B are equal.

Putting A = B in addition formula 1 gives:

$$\sin(A + A) \equiv \sin A \cos A + \cos A \sin A$$

which simplifies to:

$$\sin 2A \equiv 2 \sin A \cos A$$

This identity holds true for any case where the angle on the LHS is double the angle on the RHS.

So we might also use this identity in these (and other) forms:

$$\sin 4A \equiv 2 \sin 2A \cos 2A$$

$$\text{or} \quad \sin A \equiv 2 \sin \tfrac{1}{2}A \cos \tfrac{1}{2}A$$

Putting A = B in addition formula 3 gives:

$$\cos (A + A) \equiv \cos A \cos A - \sin A \sin A$$

which simplifies to:

$$\cos 2A \equiv \cos^2 A - \sin^2 A$$

This formula is more useful when the RHS is expressed in terms of sin A only, or cos A only.

Using Identity 2, which states that $\sin^2 A + \cos^2 A \equiv 1$, we can write:

$$\cos 2A \equiv \cos^2 A - (1 - \cos^2 A)$$

which simplifies to

$$\cos 2A \equiv 2 \cos^2 A - 1$$

Using Identity 2 again, we can write:

$$\cos 2A \equiv (1 - \sin^2 A) - \sin^2 A$$

which simplifies to

$$\cos 2A \equiv 1 - 2 \sin^2 A$$

Note again that these identities can also be used in equivalent forms such as:

$$\cos \theta \equiv 2 \cos^2(\theta/2) - 1$$

$$\text{and} \quad \cos 4x \equiv 1 - 2 \sin^2 2x \quad \text{etc.}$$

Lastly, the double angle formula for tan is derived from addition formula 5, with A = B:

$$\tan 2A \equiv \frac{2\tan A}{1-\tan^2 A}$$

It is not realistic to expect to memorise all these identities at first. The best way to learn them is by using them in exercises—proving identities and solving equations.

Proving identities

Here are two examples where we use the basic trig identities to prove other identities:

Examples (10.3)

1. Prove the identity $\dfrac{\sin^2\theta}{1-\cos\theta} \equiv 1+\cos\theta$

The best approach is to start from one side and manipulate the expression so that it equals the expression on the other side. It is best to start with the expression that can be rearranged or simplified. In this example, we choose to start with the LHS because there is nothing useful we can do to the RHS.

$$\text{LHS} = \frac{\sin^2\theta}{1-\cos\theta} \equiv \frac{1-\cos^2\theta}{1-\cos\theta} \quad \text{using Identity 2}$$

$$\text{LHS} \equiv \frac{(1-\cos\theta)(1+\cos\theta)}{1-\cos\theta} \quad \text{factorising}$$

We can cancel the term $(1 - \cos\theta)$ from the top and bottom of the fraction, provided that this term is not zero.

Note – that we have made the tacit assumption that $1 - \cos\theta \neq 0$ since this is the denominator of the LHS, and the expression would be undefined if the denominator were zero.

Cancelling $(1 - \cos\theta)$ gives $\text{LHS} \equiv 1 + \cos\theta$

Hence $\text{LHS} \equiv \text{RHS}$

2. Prove the identity $\tan^2\theta \equiv 1 - \dfrac{\cos 2\theta}{\cos^2\theta}$

Here we will start from the RHS as this has more potential for manipulation than the LHS.

First, use the basic fraction method "common denominator":

$$\text{RHS} \equiv \frac{\cos^2\theta - \cos 2\theta}{\cos^2\theta}$$

Now we can use the double angle formula for cos 2θ, to express all of the RHS in terms of cos θ:

$$\text{RHS} \equiv \frac{\cos^2\theta - (2\cos^2\theta - 1)}{\cos^2\theta}$$

$$\text{RHS} \equiv \frac{1 - \cos^2\theta}{\cos^2\theta}$$

$$\text{RHS} \equiv \frac{\sin^2\theta}{\cos^2\theta} \qquad \text{using Identity 2}$$

$$\text{RHS} \equiv \tan^2\theta \qquad \text{using Identiy 1}$$

Hence RHS ≡ LHS

Practice Questions (10.3)

Using Identity 1, Identity 2, and the double angle formulae, prove these identities.

If you get stuck, start again from the other side.

1. $(\sin\theta + \cos\theta)^2 \equiv 1 + \sin 2\theta$
2. $\sin\theta + \cos\theta \equiv \dfrac{1 - 2\cos^2\theta}{\sin\theta - \cos\theta}$
3. $2\cos^2\theta - \cos 2\theta \equiv 1$
4. $\tan 2\theta \equiv \dfrac{2\sin\theta\cos\theta}{\cos^2\theta - \sin^2\theta}$

Using identities to help solve equations

When a trig equation is given in a form in which we cannot solve it, we can often use the standard identities to manipulate the equation into a solvable form.

Examples (10.4)

1. Solve the equation: $\cos^2\theta - \sin^2\theta = 0.5$ giving solutions in the range 0° to 360°.

   We often want to simplify a trig equation into a form where only one trig function appears.

   This equation contains both sin and cos.

   But we recognise the LHS from the double angle formula $\cos 2\theta \equiv \cos^2\theta - \sin^2\theta$.

   So we write the equation as: $\cos 2\theta = 0.5$

   Now we can use the methods we applied to the equations we solved in Chapter 9:

$$2\theta = 60°, 300°, 420°, 660°$$

$$\theta = 30°, 150°, 210°, 330°$$

2. Solve the equation $\cos 2\theta + \cos\theta = 0$ giving solutions in the range 0° to 360°.

   Here we can use the double angle formula

$$\cos 2\theta \equiv 2\cos^2\theta - 1$$

and we choose this one because the equation will then only contain terms in $\cos\theta$:

$$2\cos^2\theta - 1 + \cos\theta = 0$$

This equation contains a term in $\cos^2\theta$, a term in $\cos\theta$, and a constant.

In fact, it is a quadratic equation in which the variable is $\cos\theta$.

We can write c in place of $\cos\theta$, and put the terms in the appropriate order:

$$2c^2 + c - 1 = 0$$

Then we factorise: $(2c - 1)(c + 1) = 0$

And the two solutions are: $c = ½$ and $c = -1$

i.e. $\cos\theta = ½$ and $\cos\theta = -1$

Now we return to the methods shown in Chapter 9, and we get these solutions:

$$\theta = 60°, 300°, 180°.$$

3. Solve the equation $3 \cos \theta + 2 \sin^2\theta - 3 = 0$ giving solutions in the range −180° to 180°.

Here we can use Identity 2 to replace $\sin^2\theta$, so that the equation will contain only terms in $\cos \theta$.

We write Identity 2 in the form $\sin^2\theta \equiv 1 - \cos^2\theta$ and then substitute for $\sin^2\theta$:

$$3 \cos \theta + 2(1 - \cos^2\theta) - 3 = 0$$

$$3 \cos \theta + 2 - 2 \cos^2\theta - 3 = 0$$

Now we can arrange the equation into quadratic form:

$$2 \cos^2\theta - 3 \cos \theta + 1 = 0$$

Write c for $\cos \theta$: $2c^2 - 3c + 1 = 0$

Factorise: $(2c - 1)(c - 1) = 0$

And the two solutions are $c = ½$ and $c = 1$

i.e. $\cos \theta = ½$ and $\cos \theta = 1$

Solutions in the required range are $\theta = 60°, -60°, 0°$.

4. Solve the equation $\tan 2\theta + \tan \theta = 0$ giving solutions in radians in the range 0 to 2π.

Here we write the equation in the form

$$\tan 2\theta = -\tan \theta$$

and now we can use the double angle formula for tan:

$$\frac{2\tan\theta}{1-\tan^2\theta} = -\tan\theta$$

The important point to notice here is that $\tan \theta$ is a factor on each side of the equation.

If $\theta = 0$, then $\tan \theta = 0$, and each side of the equation is equal to 0.

Hence $\theta = 0$ is one solution to the equation. It is important not to overlook this solution.

But if $\theta \neq 0$, we can divide both sides by $\tan \theta$, and we get the equation:

$$\frac{2}{1-\tan^2\theta} = -1$$

Rearranging,

$$2 = \tan^2\theta - 1$$

Then $\tan^2\theta = 3$

So $\tan\theta = \pm\sqrt{3}$

Now we have two equations,

$$\tan\theta = \sqrt{3} \quad \text{and} \quad \tan\theta = -\sqrt{3}$$

The first equation gives the solutions

$$\theta = \pi/3 \quad \text{and} \quad \theta = -2\pi/3$$

The second equation gives the solutions

$$\theta = -\pi/3 \quad \text{and} \quad \theta = 2\pi/3$$

So the complete list of solutions to the equation $\tan 2\theta + \tan\theta = 0$ is:

$$\theta = -2\pi/3, -\pi/3, 0, \pi/3, 2\pi/3$$

Radian solutions in terms of π

Notice that we gave the solutions to the last example as fractions of π. By doing this, we are writing down the exact answers, not converting to decimals and then having to round the answers. Therefore, when working in radians, and the solutions to a trig equation are simple fractions of π, always leave π in the answer. "Simple fractions of π" are the radian equivalents of angles 30°, 45°, 60°, 90° etc., and you will be able to find these solutions without using a calculator and resorting to decimals, if you have learnt the sin, cos and tan values of "Three Special Angles" which we met in Chapter 3.

Practice Questions (10.4)

In all these equations, work in radians and give solutions in the range $-\pi$ to π.

1. Use Identity 1 to help solve these equations; no other formulae are needed:
   (i) $\sin\theta = 2\cos\theta$
   (ii) $2\cos\theta + 5\sin\theta = 0$
   (iii) $\sin 2\theta - 5\cos 2\theta = 0$

2. Use Identity 2 to arrange these equations into quadratic form, and then solve them:
   (i) $3 \sin^2\theta + 2 \cos\theta = 2$
   (ii) $\cos^2\theta + 2 \sin\theta = 2$
   (iii) $4 \cos^2\theta - 4 \sin\theta - 5 = 0$
3. Use the double angle formulae to help solve these equations:
   (i) $\cos^2\theta - \sin^2\theta = -1$
   (ii) $\cos 2\theta = \sin\theta$
   (iii) $\tan 2\theta + 2 \tan\theta = 0$

Tutorial

Progress Questions (10)

In these problems you may need to use any of the identities we have met in this Chapter.

1. Prove these identities:
   (i) $\sin 4\theta \equiv 4 \sin\theta \cos\theta - 16\sin^2\theta \cos^2\theta$
   (ii) $\cos (A + B) + \cos (A - B) \equiv 2 \cos A \cos B$
   (iii) $\sin (A + B) + \sin (A - B) \equiv 2 \sin A \cos B$
2. (i) Express $\sin 3\theta$ in terms of $\sin\theta$ only
   (Hint: start with $\sin 3\theta \equiv \sin (2\theta + \theta)$ and use Addition Formula 1)
   (ii) Express $\cos 3\theta$ in terms of $\cos\theta$ only
3. Solve these equations, working in radians and giving the general solution in each case:
   (i) $\sin^2\theta + 2 \cos^2\theta = 2$
   (ii) $\sin 2\theta = \sin^2\theta - \cos^2\theta$
   (iii) $\dfrac{\tan\theta}{\tan 2\theta} = 0.25$
   (iv) $\cos (\theta + \pi) = \sin (\theta + \pi)$
   (v) $\dfrac{4\tan\theta}{\tan 2\theta} = \tan\theta - 1$

Practical Assignment (10)

Using large sheets of paper and brightly coloured pens, write down in large letters, all the identities we have met in this Chapter, and hang the posters on your wall!

Seminar discussion

Discuss the importance of distinguishing between identities and equations.

Study tip

Copy down the identities from your posters two or three times and then test yourself to see how many you can remember. [It may not be necessary to have them all memorized, as some exam boards provide lists of formulae in exams; however, the more you can remember these formulae, the more likely it is that you will become fluent in their use].

11 Turning things upside down

One-minute overview

In this chapter we investigate the functions cosec, sec, and cot. These functions are very closely related to the sin,cos and tan functions but have different properties. We will learn the definitions of these functions and look at their graphs. Getting familiar with these functions extends the range of identities and equations we can work with.

Taking a reciprocal

A "reciprocal" is an "upside down" version of some quantity. Taking a simple example, $1/x$ is the reciprocal of x (and x is the reciprocal of $1/x$).

The functions cosecant, secant and cotangent are the reciprocals of sin, cos and tan respectively.

The cosecant function

We start with the cosecant function, which is generally abbreviated to cosec.

The definition of this function is:

$$\text{cosec}\ x \equiv 1/\sin x$$

We will write a table of values of x in the first quadrant and work out the values of cosec x from sin x:

x	$\sin x$	$\text{cosec}\ x$
0°	0	1/0 (undefined)
30°	½ = 0.5	2
45°	$1\sqrt{2} = 0.707$	$\sqrt{2} = 1.414$
60°	$\sqrt{3}/2 = 0.866$	$2/\sqrt{3} = 1.155$
90°	1	1

Note that there is no function value for cosec x when $x = 0°$.

For values of x close to 0°, the values of cosec x are large, but decrease as x increases:

(Values are given here correct to 4 significant figures or more).

x	sin x	cosec x
1°	0.01745	57.23
2°	0.03489	28.65
5°	0.08716	11.47
10°	0.1736	5.759

For values of x approaching 90°, the values of cosec x decrease, approaching 1:

x	sin x	cosec x
85°	0.9962	1.0038
87°	0.9986	1.0014
89°	0.9998	1.00015

Now we have enough values to see the shape of the graph of cosec x in the first quadrant, shown in Fig 11.1:

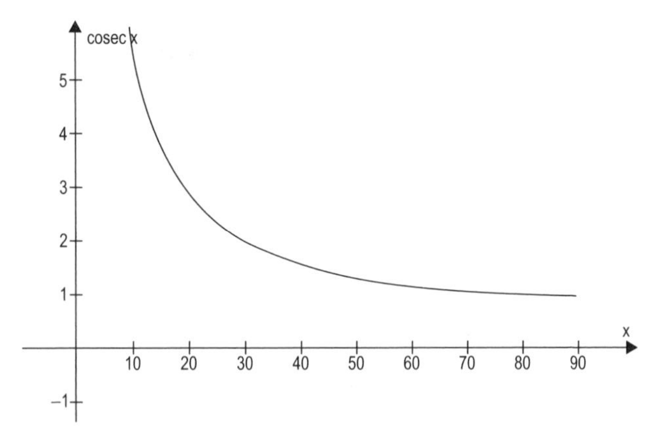

Fig 11.1

In the second quadrant, we know that sin 91° = sin 89°, sin 95° = sin 85° and so on.

This symmetry will of course also apply to the cosec function.

In the 3rd and 4th quadrants, the values of the sin function are negative; this is also true of the cosec function.

So now that we have seen the shape of the cosec curve in the first quadrant, we can extend the graph of the function to the range 0° to 360°:

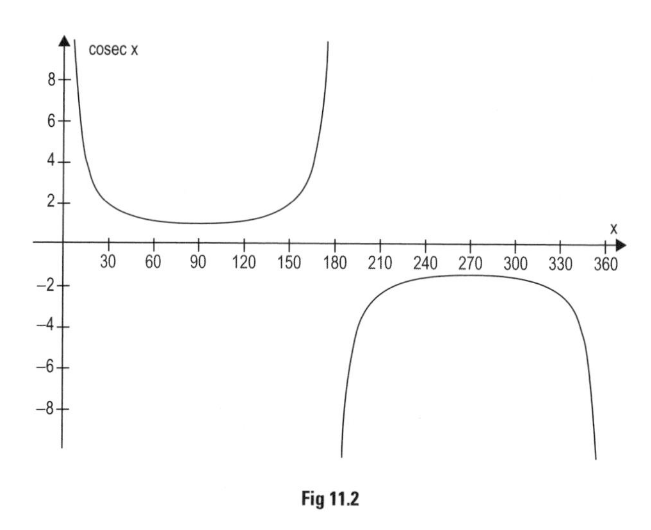

Fig 11.2

The secant function

The definition of this function (generally abbreviated to sec) is:

$$\sec x \equiv 1/\cos x$$

The cotangent function

The definition of this function (generally abbreviated to cot) is:

$$\cot x \equiv 1/\tan x$$

Practice Questions (11.1)

1. Using the same procedure as shown above for cosec x, write out tables of values for sec x, and sketch the function in the range 0° to 360°.

2. Repeat question 1, for the function cot x.
3. Complete these statements:
   (i) The minimum positive value of cosec x is …
   (ii) The maximum negative value of cosec x is …
   (iii) The minimum positive value of sec x is …
   (iv) The maximum negative value of sec x is …
4. Sketch the graphs of the two functions tan x and cot x on the same axes, in the range 0° to 90°.

   Draw in the vertical line of symmetry through the point where the curves intersect, and state the value of x at this point.

Some identities for the reciprocal trig functions

Identity 1 from Chapter 10 is:

$$\tan x \equiv \frac{\sin x}{\cos x}$$

Taking reciprocals on each side gives:

$$\cot x \equiv \frac{\cos x}{\sin x}$$ ....Identity 3

Identity 2 from Chapter 10 is:

$$\sin^2 x + \cos^2 x \equiv 1$$

Dividing each term by $\sin^2 x$ gives:

$$\frac{\sin^2 x}{\sin^2 x} + \frac{\cos^2 x}{\sin^2 x} \equiv \frac{1}{\sin^2 x}$$

which simplifies to

$$1 + \cot^2 x \equiv \operatorname{cosec}^2 x$$ ....Identity 4

Or, dividing each term by $\cos^2 x$ gives:

$$\frac{\sin^2 x}{\cos^2 x} + \frac{\cos^2 x}{\cos^2 x} \equiv \frac{1}{\cos^2 x}$$

which simplifies to

$$\tan^2 x + 1 \equiv \sec^2 x$$ ....Identity 5

Examples (11.1)

We can use these new identities, in combination with the identities we met in Chapter 10, to prove other identities. Here are two examples.

1. Prove that $2\ \text{cosec}\ 2A \equiv \sec A\ \text{cosec}\ A$

First we want to replace the reciprocal function cosec by sin:

$$\begin{aligned}
\text{LHS} &= 2\,\text{cosec}\ 2A \\
&\equiv \frac{2}{\sin 2A} \\
&\equiv \frac{2}{2\sin A\cos A} \quad \text{(using the double angle formula for sin)} \\
&\equiv \frac{1}{\sin A\cos A} \\
&\equiv \frac{1}{\sin A}\times\frac{1}{\cos A} \equiv \text{cosec}\,A\sec A = \text{RHS.}
\end{aligned}$$

2. Prove that $\sec^2\theta + \text{cosec}^2\theta \equiv \sec^2\theta\ \text{cosec}^2\theta$

This may look surprising! But we follow the same method. Start with the LHS and get rid of the reciprocal functions:

$$\begin{aligned}
\text{LHS} &= \sec^2\theta + \text{cosec}^2\theta \\
&\equiv \frac{1}{\cos^2\theta}+\frac{1}{\sin^2\theta} \\
&\equiv \frac{\sin^2\theta+\cos^2\theta}{\cos^2\theta\sin^2\theta} \quad \text{(addition of fractions using common denominator)} \\
&\equiv \frac{1}{\cos^2\theta\sin^2\theta} \quad \text{(using Identity 2)} \\
&\equiv \frac{1}{\cos^2\theta}\times\frac{1}{\sin^2\theta} \equiv \sec^2\theta\,\text{cosec}^2\theta = \text{RHS.}
\end{aligned}$$

Practice Questions (11.1)

Prove these identities:

1. $\sin\theta \tan\theta \equiv \sec\theta - \cos\theta$
2. $\cos\theta\cot\theta \equiv \dfrac{\sin\theta}{\tan^2\theta}$
3. $\text{cosec}^2\theta - \sec^2\theta \equiv \cot^2\theta - \tan^2\theta$
4. $\dfrac{\cos\theta}{1-\sin\theta} \equiv \sec\theta + \tan\theta$

Equations involving the reciprocal functions

The techniques we met for solving equations in Chapter 9 can be applied to equations which include any of the reciprocal trig functions.

Examples (11.2)

1. Solve the equation $5\,\text{cosec}\,2\theta = 12$ giving solutions in the range 0° to 360°.

   Rearrange $\text{cosec}\,2\theta = 2.4$

   Take reciprocals $\sin 2\theta = 0.417$ (to 3 d.p.)

   $2\theta = 24.6°, 155.4°$

2. Solve the equation $\tan^2\theta = \sec\theta + 1$ giving solutions in the range 0° to 360°.

   Use Identity 5 $\sec^2\theta - 1 = \sec\theta + 1$

   Rearrange into quadratic form $\sec^2\theta - \sec\theta - 2 = 0$

   Factorise $(\sec\theta - 2)(\sec\theta + 1) = 0$

   $\sec\theta = 2, \sec\theta = -1$

   Take reciprocals $\cos\theta = 0.5, \cos\theta = -1$

   Solutions $\theta = 60°, 300°, 180°$

3. Solve the equation $\cot 2\theta = \tan\theta$ giving solutions in the range 0 to 2π.

   Take reciprocal on the LHS $\dfrac{1}{\tan 2\theta} = \tan\theta$

   Use double angle formula for $\tan 2\theta$: $\dfrac{1-\tan^2\theta}{2\tan\theta} = \tan\theta$

Rearrange	$1 - \tan^2\theta = 2\tan^2\theta$
Rearrange again	$1 = 3\tan^2\theta$
Rearrange again	$\tan^2\theta = 1/3$
Take square roots	$\tan\theta = \pm 1/\sqrt{3}$
Solutions	$\theta = 0.52, 2.62, 3.67, 5.76$

Practice Questions (11.2)

Solve these equations, giving solutions in the range 0° to 360°. You may need identities from Chapter 10.

1. $\operatorname{cosec}^2\theta = 3\cot\theta - 1$
2. $\sec 2\theta = \operatorname{cosec}\theta$
3. $\operatorname{cosec} 2\theta = \sec\theta$

Tutorial

Progress Questions (11)

1. Identify which of these statements can never be true:
   (i) $\text{cosec } 2\theta = 0$
   (ii) $\sec \theta = 0.5$
   (iii) $\text{cosec } \theta + \sec \theta = 0$
   (iv) $\text{cosec}^2\theta - \cot^2\theta = -1$
2. Prove the identity: $\text{cosec } \theta - \sin \theta \equiv \cot \theta \cos \theta$
3. Solve these equations, giving solutions in the range $-\pi$ to π:
   (i) $\text{cosec } (\theta/2) = -3$
   (ii) $\tan \theta + \cot \theta = 2$
   (iii) $5 + \tan^2\theta = 3 \sec^2\theta$

Practical Assignment (11)

Use a graphical calculator to find one solution of the equation $\sec \theta = 3.5$.

Seminar discussion

Do the reciprocal functions help to increase our understanding of Trigonometry?

Study tip

Learn the definitions of $\text{cosec } \theta$, $\sec \theta$ and $\cot \theta$.

12 How to solve "Triangle" equations

One-minute overview

In this chapter we look at a particular type of trig equation which cannot be solved by methods we have learnt in earlier chapters. The method itself is not difficult, the trick is in recognising when to use it. You will need to be familiar with the addition formulae which we introduced in Chapter 10, so revise these now if you have forgotten them. We will also see that the method used to solve these equations can be used to find the maximum and minimum value of a trig expression.

What is a "triangle" equation?

A "triangle" equation is in this form: $a \sin \theta \pm b \cos \theta = c$ where a, b and c are constants.

Note that *this is not a standard name* for this type of equation, but it is a good one to use, since the method we use to solve equations like this one involves sketching a right-angled triangle. The correct mathematical name for this technique is "Harmonic form".

Examples (12.1)

1. Look at this equation: $4 \sin \theta + 3 \cos \theta = 1$

   It doesn't take long to see that we cannot solve this equation using the identities we have met.

   It is not possible to rearrange this equation so that it only contains $\sin \theta$, or only contains $\cos \theta$.

   So we need a new method. Here is the method for this example:

Step 1: Sketch a right-angled triangle with the two shorter sides lengths 4 and 3 (these are the coefficients of $\sin \theta$ and $\cos \theta$ in the equation).

Step 2: Label one of the non-right angles α. You can choose either angle (we will see in the next example what happens if we choose the other angle).

Step 3: Use Pythagoras to find the length of the hypotenuse. In this example it is 5.

Step 4: Write down $\sin \alpha$ and $\cos \alpha$ from the triangle:

$$\sin\alpha = \frac{3}{5} \qquad \cos\alpha = \frac{4}{5}$$

Step 5: Find the value of the angle α:

$$\alpha = \arcsin(3/5) = 36.9°$$

Step 6: Divide every term in the equation by the length of the hypotenuse:

$$\frac{4}{5}\sin\theta + \frac{3}{5}\cos\theta = 0.2$$

Step 7: Replace the fractions in the equation with $\sin \alpha$ and $\cos \alpha$:

$$\cos\alpha \sin\theta + \sin\alpha \cos\theta = 0.2$$

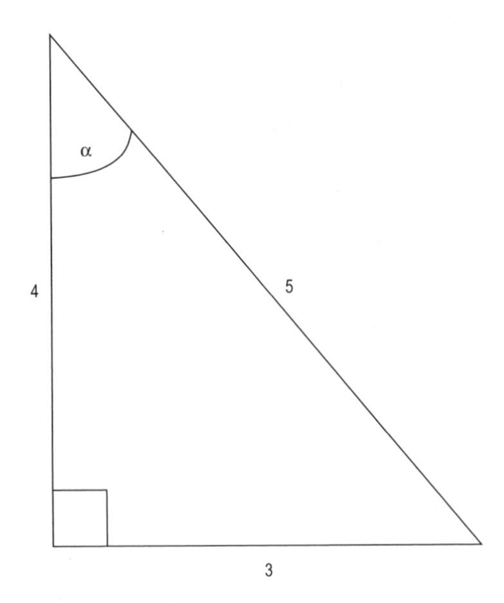

Fig 12.1

Now we come to the central point of this method. To do the next step, you need to be able to choose the appropriate one of the first four Addition Formulae. Here they are again:

1. **$\sin(A + B) \equiv \sin A \cos B + \cos A \sin B$**
2. **$\sin(A - B) \equiv \sin A \cos B - \cos A \sin B$**
3. **$\cos(A + B) \equiv \cos A \cos B - \sin A \sin B$**
4. **$\cos(A - B) \equiv \cos A \cos B + \sin A \sin B$**

Step 8: Compare the equation with the RHS of the addition formulae and choose the formula which matches the form of the equation. In this case, it is No. 1.

If you are in any doubt about this, write the equation with the terms the other way round:

$$\sin\alpha\cos\theta + \cos\alpha\sin\theta = 0.2$$

Now you see we can write the LHS as $\sin(\theta + \alpha)$. So the equation is now

$$\sin(\theta + \alpha) = 0.2$$

The essential work has been done. We have changed the equation from its original form with two different trig functions, sin and cos, into an equation containing only the sin function.

Step 9: Using the techniques we learnt in Chapter 9, we can now complete the solution:

$$\theta + \alpha = 11.5°, 168.5°$$

$$\theta = 11.5° - \alpha, 168.5° - \alpha$$

$$\theta = 11.5° - 36.9°, 168.5° - 36.9°$$

$$\theta = -25.4°, 131.6°$$

Assuming we are required to give solutions in the range 0° to 360°, we bring the first angle into the range by adding 360°:

$$-25.4° + 360° = 334.6°$$

So the two solutions are

$$\underline{\theta = 131.6°, 334.6°}$$

2. Now let's see what happens if we make α the other angle in the triangle:

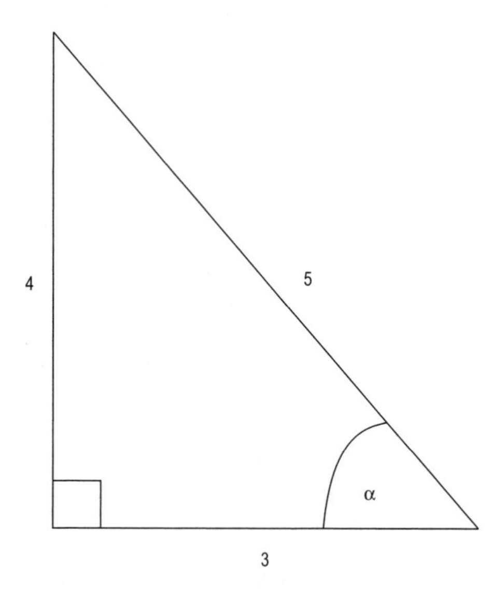

Fig 12.2

$$\sin\alpha = \frac{4}{5} \qquad \cos\alpha = \frac{3}{5} \qquad \alpha = 53.1°$$

The equation becomes:

$$\sin\alpha \sin\theta + \cos\alpha \cos\theta = 0.2$$

and this time we need addition formula No. 4, giving:

$$\cos(\theta - \alpha) = 0.2$$

$$\theta - \alpha = 78.5°, 281.5°$$

$$\theta = 78.5° + 53.1°, 281.5° + 53.1°$$

$$\underline{\theta = 131.6°, 334.6°}$$

So it's reassuring to see that either choice of angle for α leads to the same solutions!

3. Here is another equation:

$$5 \cos \theta - 2 \sin \theta = 3$$

Note that we take the values 5 and 2 for the sides of the triangle. The method is not affected by the minus sign in the equation. Following the steps in the same order as before, we have:

Hypotenuse $= \sqrt{(5^2 + 2^2)} = \sqrt{29}$.

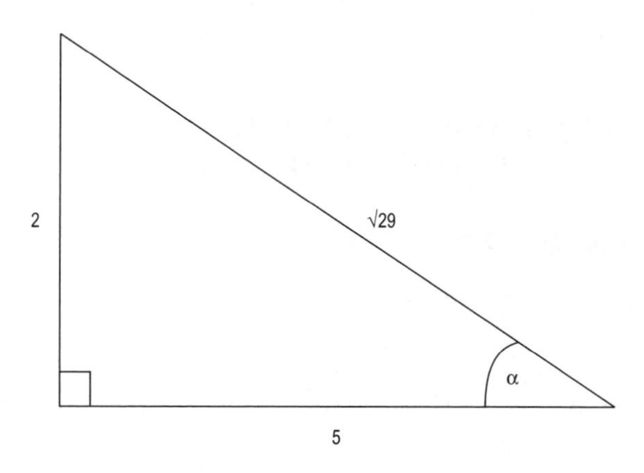

Fig 12.3

We leave $\sqrt{29}$ in this form, until we reach the point of working out the answers.

$$\cos\alpha = \frac{5}{\sqrt{29}} \qquad \sin\alpha = \frac{2}{\sqrt{29}}$$

$$\alpha = 21.8°$$

$$\frac{5}{\sqrt{29}}\cos\theta - \frac{2}{\sqrt{29}}\sin\theta = \frac{3}{\sqrt{29}}$$

$\cos \alpha \cos \theta - \sin \alpha \sin \theta = 0.557$

Using Addition Formula No. 3:

$$\cos(\alpha + \theta) = 0.557$$

$$\alpha + \theta = 56.1°, 303.9°$$

$$\theta = 56.1° - 21.8°, 303.9° - 21.8°$$

$$\underline{\theta = 34.3°, 282.1°}$$

Practice Questions (12.1)

1. Solve the equation $5 \cos \theta - 2 \sin \theta = 3$ (Example 3 above), with the angle α in the other corner of the triangle, so that $\cos\alpha = 2/\sqrt{29}$ and $\sin\alpha = 5/\sqrt{29}$. Check your answers are the same as above.
2. Solve these equations, giving all solutions in the range 0° to 360°:

   (i) $\sin \theta + 2 \cos \theta = 1$ (ii) $3 \cos \theta - 2 \sin \theta = 1.5$

   (iii) $3 \sin \theta - \cos \theta = 2$ (iv) $\sqrt{2} \cos \theta + \sin \theta = 1.2$

Using Harmonic form to find maximum and minimum values

Examples (12.2)

1. Let's take the LHS of our first equation above:

$$4 \sin \theta + 3 \cos \theta$$

This expression includes two trig functions, and can itself form a function:

$$y = 4 \sin \theta + 3 \cos \theta$$

We can plot the graph of this function, and it is shown in Fig 12.4. [Try this on a graphical calculator.]

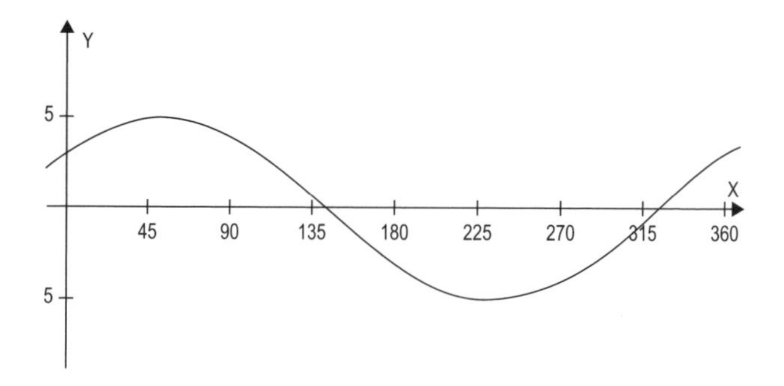

Fig 12.4

Now let's take the function:

$$y = 4 \sin \theta + 3 \cos \theta$$

and apply almost the same method as that which we used to solve the equation $4 \sin \theta + 3 \cos \theta = 1$.

Steps 1 to 5 are just the same as before, so we have:

$$\sin\alpha = \frac{3}{5} \qquad \cos\alpha = \frac{4}{5} \qquad \alpha = 36.9^0$$

Step 6 is a bit different. We don't have an equation, so we can't divide the RHS by 5.

Instead, we divide each term by 5, and write 5 outside a bracket, so that we do not change the value of the function:

$$y = 5\left(\frac{4}{5}\sin\theta + \frac{3}{5}\cos\theta\right)$$

Now we can replace the fractions with $\sin \alpha$ and $\cos \alpha$, as before:

$$y = 5(\cos \alpha \sin \theta + \sin \alpha \cos \theta)$$

and again use Addition Formula 1:

$$y = 5 \sin (\theta + \alpha)$$

All we have done is transform the function $y = 4 \sin \theta + 3 \cos \theta$ into the different form $y = 5 \sin (\theta + \alpha)$. Apart from Step 6, the method is the same. You will need this method if you are asked to find the maximum or minimum value of an expression in the form $a \sin \theta + b \cos \theta$. You may also be asked to find the value of θ at which the maximum or minimum occurs. Note that this use of the words "maximum" and "minimum" has nothing to do with the calculus methods of finding maximum and minimum values (though indeed it is possible to find these values by using calculus). Sometimes the words "least" and "greatest" values are used.

In Chapter 8, we saw that the graph of the function $y = \sin (\theta + \alpha)$ is a translation of the graph of the function $y = \sin \theta$, α degrees to the left. And we saw that the graph of the function $y = 5 \sin \theta$ is a vertical stretch of the graph of $y = \sin \theta$, with all the y-values multiplied by 5. In this example, with $\alpha = 36.9°$, the graph of the function $y = \sin (\theta + \alpha)$ will cut the x-axis at $(180°–36.9°)$. Look again at Fig 12.4, or use your graphical calculator to produce the graph of the function $y = 5 \sin(\theta + 36.9°)$. You will see that the graphs of $y = 4 \sin \theta + 3 \cos \theta$ and $y = 5 \sin(\theta + 36.9°)$ are the same. This just confirms that these are two different forms of the same function.

In our example, the maximum (or greatest) value of the function $4 \sin \theta + 3 \cos \theta$ is 5, and the minimum (or least) value is –5. You can see this from the graph in Fig 12.4. But you do not need to draw a graph each time. The maximum and minimum values are always R and –R, where R is the length of the hypotenuse.

What are the values of θ when the function has its maximum or minimum values? We have two simple equations to solve to answer this question:

$$5 \sin(\theta + \alpha) = 5 \quad \text{and} \quad 5 \sin(\theta + \alpha) = -5$$

The first equation gives

$$\sin(\theta + \alpha) = 1$$

$$\theta + \alpha = 90°$$

$$\theta = 90° - 36.9°$$

$$\theta = 53.1°$$

The second equation gives

$$\sin(\theta + \alpha) = -1$$

$$\theta + \alpha = 270°$$

$$\theta = 270° - 36.9°$$

$$\theta = 233.1°$$

Look again at Fig 12.4 and you will see that the highest part of the curve occurs just to the right of 45° and the lowest part of the curve occurs just to the right of 225°, which is consistent with these results.

If a question specifies exactly the form required for the answer, we have to be careful to make the correct choice for the angle α in the triangle, as in the next example.

2. (i) Express $5\cos\theta - 2\sin\theta$ in the form $R\sin(\theta - a)$
   (ii) State the greatest and least values of the expression $5\cos\theta - 2\sin\theta$
   (iii) Find the values of θ corresponding to the greatest and least values.

(i) We have already drawn a triangle for this expression, in Fig 12.3 above, when we were solving the equation $5\cos\theta - 2\sin\theta = 3$. The hypotenuse is $\sqrt{29}$.

Dividing by $\sqrt{29}$ and taking $\sqrt{29}$ outside a bracket gives:

$$\sqrt{29}\left(\frac{5}{\sqrt{29}}\cos\theta - \frac{2}{\sqrt{29}}\sin\theta\right)$$

Notice that we want to arrive at $\sin(\theta - \alpha)$, so we will have to use Addition Formula 2.

This means we must get the expression into the form:

$$\sin\alpha\cos\theta - \cos\alpha\sin\theta$$

So we must have $\sin\alpha = 5/\sqrt{29}$ and $\cos\alpha = 2/\sqrt{29}$

and this means that the angle α will not be as in Fig 12.3, but will be in the other corner, as shown in Fig 12.5:

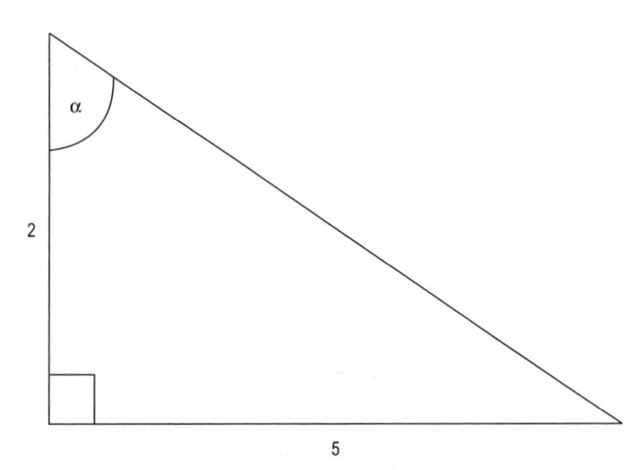

Fig 12.5

So now the expression becomes:

$$\sqrt{29}(\sin \alpha \cos \theta - \cos \alpha \sin \theta) = \sqrt{29} \sin (\alpha - \theta)$$

Using the fact that $\sin (-\theta) = \sin \theta$, this becomes:

$$-\sqrt{29} \sin (\theta - \alpha)$$

but the minus sign does not affect the greatest and least values.

We have $R = \sqrt{29}$, and $\alpha = \text{arc} \sin (5/\sqrt{29}) = 68.2°$

(ii) The greatest and least values of

$$5 \cos \theta - 2 \sin \theta \text{ are } \sqrt{29} \text{ and } - \sqrt{29}$$

(iii) The value of θ when $5 \cos \theta - 2 \sin \theta = \sqrt{29}$ is given by the equation

$$\sqrt{29} \sin (\theta - a) = \sqrt{29}$$

$$\text{i.e.} \quad \sin (\theta - \alpha) = 1$$

$$\theta - \alpha = 90°$$

$$\theta = 90° + \alpha = 158.2°$$

and the value of θ when $5 \cos \theta - 2 \sin \theta = -\sqrt{29}$ is given by the equation

$$\sqrt{29} \sin (\theta - a) = -\sqrt{29}$$

$$\text{i.e.} \quad \sin (\theta - \alpha) = -1$$

$$\theta - \alpha = 270°$$

$$\theta = 338.2°$$

Practice Questions (12.2)

For each of the following functions,

(i) express the function in the form $R \sin (\theta \pm \alpha)$

(ii) write down the greatest and least values of the function

(iii) find the values of θ at which the greatest and least values occur

1. $y = 2 \sin \theta - 4 \cos \theta$
2. $y = 3 \sin \theta + \cos \theta$
3. $y = 4 \sin \theta + 5 \cos \theta$

Tutorial

Progress Questions (12)

1. (i) Sketch the graph of the function $y = 2\sin(\theta + 30°)$ in the range 0° to 360°.

   (Start with the graph of $y = \sin\theta$, then translate 30° to the left and do a vertical stretch by factor 2).

   (ii) Write down the maximum and minimum values of this function.

   (iii) Write down the value of θ at which the function has its maximum value.

   (iv) Write down the value of θ at which the function has its minimum value.

   (v) Read, from your graph, two values of θ for which $y = 1$.

   (vi) Apply Addition Formula 1 to the function $y = 2\sin(\theta + 30°)$ and replace cos 30° and sin 30° by their fraction values (See Three special Angles in Chapter 3 if you need help here) and simplify your answer.

   (vii) Write down two solutions to the equation $\sqrt{3}\sin\theta + \cos\theta = 1$ (no further work should be necessary).
2. Solve the equation $5\cos\theta - 7\sin\theta = 3$, working in radians and giving solutions in the range $-\pi$ to π.
3. Find the greatest and least values of the function $y = 5\sin\theta + 12\cos\theta$.

Practical Assignment (12)

Using a graphical calculator:

(i) plot the function $y = 2\sin(\theta + 30°)$.

(ii) solve the equation $\sqrt{3}\sin\theta + \cos\theta = 0.8$, giving solutions in the range 0° to 360°.

Note – this is the same function as in Progress Questions (12) 1, but this time you are looking for values of θ for which $y = 0.8$.

Seminar discussion

If the method of harmonic form had not been developed, how might you try to solve an equation of the form $a \sin \theta \pm b \cos \theta = c$?

Study tip

Read again through the first example in this Chapter, then close the book and write down the steps from memory.

13 How to solve equations which have two different angles

One-minute overview

In this chapter we look another type of trig equation which cannot be solved by methods we have learnt before. We introduce the "factor formulae", which are derived from the Addition Formulae, and we see how to use these to solve equations which involve two different angles. Then we extend the use of the Factor Formulae to proving identities.

Equations with two different angles

In Chapter 10, Examples 10.4, we solved this equation:

$$\cos 2\theta + \cos \theta = 0$$

We used the double-angle formula for $\cos 2\theta$ and converted the equation into a quadratic in $\cos \theta$.

So we were able to change the equation from one with two different angles 2θ and θ, into an equation with only one angle, θ.

Could we apply a similar method to this equation?

$$\cos 3\theta + \cos \theta = 0$$

The answer is a cautious yes, because there is an identity which enables us to express $\cos 3\theta$ in terms of θ.

But this identity contains a term in $\cos^3\theta$. So we would have a cubic equation, and if we could not factorise it then we would be stuck! In any case, it soon becomes obvious that this type of method will not work when we are looking at equations like $\cos 5\theta + \cos 7\theta = 0$ or $\sin 6\theta - \sin 4\theta = 0$.

Finding a new method

The Addition Formulae come in handy yet again. Addition Formulae 1 and 2 are:

$$\sin (A + B) \equiv \sin A \cos B + \cos A \sin B$$

$$\sin (A - B) \equiv \sin A \cos B - \cos A \sin B$$

Adding these together, we get:

$$\sin (A + B) + \sin (A - B) \equiv 2 \sin A \cos B$$

Now we make a substitution so that we have single angles on the LHS:

Let $C = A + B$ and $D = A - B$. Rearranging these two equations gives $A = \tfrac{1}{2}(C + D)$ and $B = \tfrac{1}{2}(C - D)$.

So our new identity is:

$$\sin C + \sin D \equiv 2 \sin \tfrac{1}{2}(C + D) \cos \tfrac{1}{2}(C - D)$$

Before we see how to use this identity to solve equations, we need to complete this new set of identities, and this is set as an exercise for you in the following questions. You will find it useful to remember that $\cos (-\theta) = \cos \theta$ and $\sin (-\theta) = -\sin \theta$.

Practice Questions (13.1)

1. Write down Addition Formulae 1 and 2, as above. Subtract formula 2 from formula 1, and using C and D as above, derive the identity $\sin C - \sin D \equiv 2 \cos \tfrac{1}{2}(C + D) \sin \tfrac{1}{2}(C - D)$.
2. Write down Addition Formulae 3 and 4. Refer back to Chapter 10 if you have forgotten these.

   Add the two formulae together, and derive the identity $\cos C + \cos D \equiv 2 \cos \tfrac{1}{2}(C + D) \cos \tfrac{1}{2}(C - D)$.
3. Write down Addition Formulae 3 and 4 again. Subtract formula 4 from formula 3, and derive the identity $\cos C - \cos D \equiv -2 \sin \tfrac{1}{2}(C + D) \sin \tfrac{1}{2}(C - D)$.

Factor formulae

The four formulae we have now derived are called the Factor Formulae. This is because they allow us to express the sum or difference of two sines or cosines in a form containing two factors. This is important because we will be able to rearrange an equation such as $\cos 3\theta + \cos \theta = 0$ and then factorise it.

Here are the factor formulae so that you can refer to them while working through the rest of this chapter:

1. $\sin C + \sin D \equiv 2 \sin \frac{1}{2}(C + D) \cos \frac{1}{2}(C - D)$
2. $\sin C - \sin D \equiv 2 \cos \frac{1}{2}(C + D) \sin \frac{1}{2}(C - D)$
3. $\cos C + \cos D \equiv 2 \cos \frac{1}{2}(C + D) \cos \frac{1}{2}(C - D)$
4. $\cos C - \cos D \equiv -2 \sin \frac{1}{2}(C + D) \sin \frac{1}{2}(C - D)$

Practice Questions (13.2)

1. Use the factor formulae to rearrange these expressions in factor form:
   (i) $\sin 40° + \sin 30°$ (ii) $\sin 50° - \sin 30°$
   (iii) $\cos 60° + \cos 20°$ (iv) $\cos 80° - \cos 20°$
2. Using your calculator, find the value of the LHS and the RHS of each identity in question 1, and verify that they are equal.
3. Rearrange these expressions in factor form:
   (i) $\sin 2\theta + \sin 4\theta$ (ii) $\cos 2\theta + \cos 4\theta$
   (iii) $\cos 3\theta - \cos\theta$ (iv) $\sin 5\theta - \sin 3\theta$

Using the factor formulae to solve equations

Example (13.1)

Let's return to the equation

$$\cos 2\theta + \cos \theta = 0$$

Applying Factor Formula 3 gives

$$2 \cos (3\theta/2) \cos \theta/2 = 0$$

Ignoring the factor 2, we have two factors which multiply to make 0.

These are cos $(3\theta/2)$ and cos $\theta/2$. Since these multiply to make 0, one or both of these factors is 0.

If the first factor is 0, we have this equation:

$$\cos(3\theta/2) = 0$$

and the solutions to this equation are

$$3\theta/2 = 90°, 270°, 450°$$

$$\theta = 60°, 180°, 300°$$

If the second factor is 0, we have this equation:

$$\cos \theta/2 = 0$$

and the solutions to this equation are

$$\theta/2 = 90°, 270°$$

$$\theta = 180°, 540°$$

So the complete set of solutions in the range 0° to 360° is $\theta = 60°, 180°, 300°$.

And these match the solutions we found by the previous method in Chapter 10 (Examples 10.4, 2).

And you see that this new method using the factor formulae takes less work.

Practice Questions (13.3)

Use the factor formulae to solve these equations, giving solutions in the range 0° to 180°.

1. $\sin 2\theta + \sin 4\theta = 0$
2. $\cos 2\theta + \cos 4\theta = 0$
3. $\cos 3\theta - \cos \theta = 0$
4. $\sin 5\theta - \sin 3\theta = 0$

Examples (13.2)

Now we can try some slightly harder examples.

Equations with three terms can be solved using the factor formulae:

1. $\sin 3\theta + \sin \theta + \sin 2\theta = 0$

   Applying Factor Formula 1 to the first two terms, we get:

$$2 \sin 2\theta \cos \theta + \sin 2\theta = 0$$

Now we have two terms, which have the common factor $\sin 2\theta$, so we take this outside a bracket:

$$\sin 2\theta(2 \cos \theta + 1) = 0$$

Again, we apply the argument that there are two factors which multiply to 0, so one or both factors is 0.

Either

$$\sin 2\theta = 0 \quad \text{or} \quad 2 \cos \theta + 1 = 0$$

The equation

$$\sin 2\theta = 0$$

has solutions

$$2\theta = 0°, 180°, 360°, 540°, 720°$$

$$\theta = 0°, 90°, 180°, 270°, 360°$$

The equation

$$2 \cos \theta + 1 = 0$$

rearranges to

$$\cos \theta = -\tfrac{1}{2}$$

and the solutions are

$$\theta = 120°, 240°$$

So the complete set of solutions in the range 0° to 360° is $\theta = 0°, 90°, 120°, 180°, 240°, 270°, 360°$.

2. $\cos 3\theta + \sin \theta - \cos \theta = 0$

We can apply the factor formulae to $\cos 3\theta - \cos \theta$ and this gives:

$$-2 \sin 2\theta \sin \theta + \sin \theta = 0$$

Take the factor $\sin\theta$ outside a bracket:

$$\sin \theta\,(-2 \sin 2\theta + 1) = 0$$

and we get the two equations

$$\sin \theta = 0 \quad \text{and} \quad -2 \sin 2\theta + 1 = 0$$

The first equation

$$\sin \theta = 0$$

has solutions

$$\theta = 0°, 180°, 360°$$

and the second equation

$$-2 \sin 2\theta + 1 = 0$$

rearranges to

$$\sin 2\theta = ½$$

giving

$$2\theta = 30°, 150°, 390°, 510°$$

$$\theta = 15°, 75°, 195°, 255°$$

So the complete set of solutions in the range 0° to 360° is $\theta = 0°, 15°, 75°, 180°, 195°, 255°, 360°$.

Here are some examples for you to try. You have to decide which two terms to combine using the factor formula. If you make a wrong choice, you will get an equation which you can't factorise. So just return to the original equation and make a different choice. Generally, it is better to combine two angles such that $C + D$ and $C - D$ are even, not odd, so that dividing by 2 does not lead to half-angles. For example, 5θ and 3θ lead to the angles 4θ and θ, but 5θ and 4θ would lead to the angles $9\theta/2$ and $\theta/2$. And a term with a coefficient, such as $2 \cos \theta$, clearly cannot be combined with another term.

Practice Questions (13.4)

Use the factor formulae to solve these equations, giving solutions in the range 0° to 180°:

1. $\sin 3\theta - \sin \theta + \cos 2\theta = 0$
2. $\cos 3\theta + \cos 4\theta + \cos 5\theta = 0$
3. $\cos 4\theta + 2 \sin \theta - \cos 2\theta = 0$
4. $\sin 5\theta + 2 \cos 2\theta + \sin \theta = 0$

Using the factor formulae to prove identities

Examples (13.3)

1. Prove that $\dfrac{\sin 3\theta + \sin\theta}{\cos 3\theta + \cos\theta} = \tan 2\theta$

   Using the factor formulae on the LHS, we have:

$$\text{LHS} = \frac{2\sin 2\theta\cos\theta}{2\cos 2\theta\cos\theta}$$

   Cancelling out the common factors, we have:

$$\text{LHS} = \frac{\sin 2\theta}{\cos 2\theta}$$

   And referring to Identity 1, we have:

$$\text{LHS} = \tan\, 2\theta.$$

2. Prove that $\dfrac{\cos 3\theta - \cos\theta}{\cos 3\theta + \cos\theta} = -\tan\theta\tan 2\theta$

Again, we apply the factor formulae to the LHS, and we have:

$$\text{LHS} = \frac{-2\sin 2\theta\sin\theta}{2\cos 2\theta\cos\theta}$$

Cancelling out the factor 2, and referring to Identity 1, we have:

$$\text{LHS} = -\tan\, 2\theta\,\tan\,\theta$$

Practice Questions (13.5)

Prove these identities:

1. $\dfrac{\sin\theta + \sin 3\theta + \sin 5\theta + \sin 7\theta}{\cos\theta + \cos 3\theta + \cos 5\theta + \cos 7\theta} \equiv \tan 4\theta$

2. $\dfrac{\cos 4\theta + \cos 2\theta}{\sin 4\theta + \sin 2\theta} \equiv \cot 3\theta$

Tutorial

Progress Questions (13)

1. (i) Show that the factor formula

$$\cos C + \cos D \equiv 2 \sin \tfrac{1}{2}(C + D) \sin \tfrac{1}{2}(C - D)$$

can be written in the form

$$\cos A \cos B \equiv \tfrac{1}{2} [\cos (A + B) + \cos (A - B)]$$

(ii) Show that the factor formula

$$\cos C - \cos D \equiv -2 \sin \tfrac{1}{2}(C + D) \sin \tfrac{1}{2}(C - D)$$

can be written in the form

$$\sin A \sin B \equiv \tfrac{1}{2} [\cos (A - B) - \cos (A + B)]$$

(Hint: refer back to the Addition Formulae).

2. By choosing suitable values for C and D, use the factor formula:

$$\sin C - \sin D \equiv 2 \cos \tfrac{1}{2}(C + D) \sin \tfrac{1}{2}(C - D)$$

to verify that

$$\cos 45° \sin 45° = \tfrac{1}{2}$$

3. Using the results of question 1, show that:

$$\sin 3\theta \sin \theta + \cos 3\theta \cos \theta \equiv \cos 2\theta$$

4. Use the identity you have proved in question 3 to solve this equation, giving the general solution in radians:

$$\sin 3\theta \sin \theta + \cos 3\theta \cos \theta = 3 \sin 2\theta$$

Practical Assignment (13)

Without using the factor formulae, try a Trial and Improvement method to solve the equation:

$$\sin 3\theta + \sin \theta = 0$$

Seminar discussion

In what other branch of mathematics are the factor formulae useful in reverse, i.e. converting a trig expression containing factors to an expression containing a sum (or difference) of terms/

Study tip

Read the "One-minute overview" of each chapter to find out your weaker areas, and revise these before you start the Revision Examples in the next chapter.

14 Revision Questions

One-minute overview

This chapter contains miscellaneous revision questions. They are not presented in the same order as the material in the book. Therefore, this chapter gives you some practice in identifying the methods you need to apply to answer each question. This is important, because this is exactly what you need to be able to do in an exam.

1. In the triangle shown in Fig 14.1,
   a) Find the angle α
   b) Find the length of the hypotenuse.

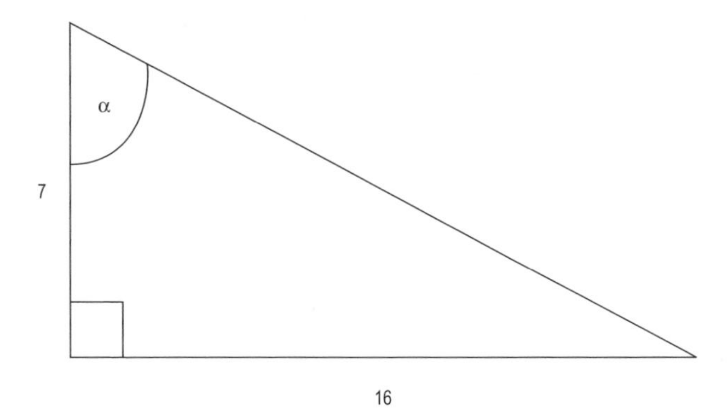

Fig 14.1

2. Sketch the triangle PQR with PQ = 7 cm, and QR = 5 cm. Find the length of PR:
   a) if $\angle PQR = 90°$
   b) if $\angle PQR = 65°$
3. Convert these directions to a bearing:
   a) NE
   b) NNW
   c) ESE

4. a) Complete this table, giving radians as fractions of π (not as decimals):

Degrees	Radians
180°	π
90°	
	$\pi/3$
45°	
270°	
	$2\pi/3$
30°	
	$5\pi/6$

b) The angles in the following table are given in radians. Complete the table, showing in which quadrant each angle belongs:

Angle	Quadrant
3.5	3
$2\pi/3$	2
0.96	
5.9	
$3\pi/4$	
1.5	
$5\pi/8$	

5. Solve these equations, working in radians and giving solutions in the range $-\pi$ to π:
   a) $\tan 2\theta = 1.4$
   b) $\cos \frac{1}{2}\theta = -1$
   c) $\sin(\theta + \pi/2) = 0.5$
6. Write down the six Addition Formulae, from memory.
7. The floor of a greenhouse is in the shape of a regular octagon. The distance straight across the floor, from one vertex of the octagon, to the opposite vertex, is 10 metres.

(i) Draw a diagram of the octagon, showing it divided up into 8 identical isosceles triangles.
(ii) Work out the sizes of the angles in each triangle.
(iii) Find the distance straight across the floor, from the mid-point of one side of the octagon to the mid-point of the opposite side.

8. Triangle FGH has $\angle FGH = 72°$, FG = 13 cm and $\angle FHG = 50°$. Find the lengths of GH and FH.
9. Prove these identities:
   a) $1 + \tan^2\theta \equiv \sec^2\theta$
   b) $\dfrac{\cot\theta}{\text{cosec}\theta} \equiv \dfrac{\cos\theta}{1-\cos^2\theta}$
10. Solve these equations, giving solutions in the range 0° to 360°:
    a) $3 \sin\theta - 7 \cos\theta = 2$
    b) $\cos 3\theta + \cos 5\theta + \cos 7\theta = 0$
    c) $2 \tan\theta = \sec^2\theta$
11. Draw an equilateral triangle, with one line of symmetry, so that the triangle is divided into two right-angled triangles. Show that:
    a) $\sin 60° = \sqrt{3}/2$
    b) $\tan 30° = 1/\sqrt{3}$
    c) $\sec 60° = 2$
12. a) Write down 3 different ways of expressing this statement mathematically:
    "The angle whose cosine is 0.4 is 66.4°."
    b) Evaluate
       (i) $\sin^2(\cos^{-1} 0.8)$
       (ii) $\sec^{-1}(\text{cosec}^2(\pi/4))$
13. Blackwood Farm is 20 km from the nearest town, on a bearing of 130°. The main road into the town runs into it from the South-East. To get from Blackwood Farm to the town, it is necessary to drive due South along a straight country lane, to get to the nearest access point to the main road.

    Calculate the distance by road from Blackwood Farm to the town, giving your answer correct to 2 d.p.
14. Express the function $y = 3 \sin\theta - \cos\theta$ in harmonic form. (There are two possible answers)

Use your result to help you draw a sketch of the function.

15. (i) Express $\tan 4\theta$ in terms of $\tan 2\theta$ only.
    (ii) Express $\tan 3\theta$ in terms of $\tan \theta$ only.
16. Sketch a cube with sides of length 1 unit. Draw in the diagonal across the base, and the diagonal from one end of this base diagonal to the opposite top vertex.

    Find (i) the length of the base diagonal
    (ii) the length of the corner-to-corner diagonal
    (iii) the angle between these two diagonals.

Answers

Practice Questions (1.1)

1. 30°   2. South   3. 18° per minute

Practice Questions (1.2)

1. $\angle AXD = 180°$ 2. $\angle AEB = 77°$ 3. $\angle DBC = 30°$

Practice Questions (1.3)

1.

65°	A
165°	O
265°	R
184°	R
101°	O
86°	A

2.

30°, 60°	C
21°, 69°	C
45°, 55°	N
20°, 160°	S
67°, 113°	S
150°, 40°	N

3. a) (i) 720° (ii) 1080° (iii) 540° b) 180°

Practice Questions (1.4)

1. (i) $\pi \approx 3.14$ (ii) $1^C < 60°$ (iii) $360° > 6^C$ (iv) $2\pi^C = 360°$
2. (i) π (ii) $\frac{1}{2}\pi$ (iii) 4π (iv) $3\pi/2$

Progress Questions (1)

1. (i) $(90 - \alpha)°$ (ii) $(180 - \alpha)°$
3. (i) Complementary angles $\angle ABC$ and $\angle BCA$
   (ii) 60° 40° 18° 73.7° 8.5°
4. 57.3° correct to 1 d.p.

Practice Questions (2.1)

1. a) 1:3 1:5 3:7 2:3   b) 0.125 0.25 0.6 0.08
2. 7.5 cm   3. 30°   4. EG = 3.2 cm HK = 10 cm KL = 14 cm

Practice Questions (2.2)

1. No. $8^2 \neq 4^2 + 6.8^2$.   2. $10^2 = 100$. $6^2 + 8^2 = 64 + 36 = 100$.
   $5^2 = 25$. $3^3 + 4^2 = 9 + 16 = 25$.
3. $h^2 = 49$. $a^2 = 49 - 3^2 = 40$. $a = 6.32$ to 2 d.p.

4.

adjacent $>$ hypotenuse	N
opposite = adjacent	Y
hypotenuse = opposite	N
$\angle\alpha = 90°$	N
opposite $<$ hypotenuse	Y
opposite + adjacent = hypotenuse	N

Progress Questions (2)

1. opposite : adjacent = 1 : 1 opposite : hypotenuse = 1 : 1.4
3. $\angle BAC = 48°$ 4. $\angle BAC = 48°$
5. opposite : hypotenuse = 1 : 2

Practical Assignment (2)

As the angle gets smaller, the opposite side gets smaller and the complementary angle gets larger.

The ratio adjacent:hypotenuse is getting closer to 1.

When the angle reaches 0°, the opposite side disappears and the ratio adjacent:hypotenuse = 1 : 1.

With the other angle as reference angle, the ratio opposite : hypotenuse gets closer to 1 as the angle gets larger, and when this angle reaches 90°, the adjacent disappears so the ratio adjacent : hypotenuse = 0.

Practice Questions (3.1)

1. 0.87 2. 0.67

Practice Questions (3.2)

1. 0.174 0.342 0.500 0.643 0.766 0.866 0.940 0.985
4. 17.5° 34.1° 41.3° 66.9° 75.9°

Practice Questions (3.3)

1. (i) 0.50 (ii) 1.73 2. (i) 0.74 (ii) 0.90
4. (i) 0.783 (ii) 38.5° 5. (i) 2.86 (ii) 0.79, 38.5°
6. 0.985 0.940 0.866 0.766 0.643 0.500 0.342 0.174
8. 72.5° 55.9° 48.7° 23.1° 14.1°

Practice Questions (3.4)

1. $1/\tan(90 - \alpha) = 1/(AC/CB) = 1 \div (AC/CB) = 1 \times CB/AC = \tan \alpha$

   $\sin \alpha/\cos \alpha = (CB/AB) \div (AC/AB) = (CB/AB) \times (AB/AC) = CB/AC = \tan \alpha$
2. (i) 0.966 (ii) 0.259 (iii) 3.73 (iv) 0.27
3. (i) 1 (ii) 0 (iii) 0 (iv) 1
5. (i) 35° (ii) 45° (iii) 65°

Progress Questions (3)

1. For any angle α in the range 0° to 90°, $\sin \alpha$ and $\cos \alpha$ are always less than 1 because the hypotenuse is always longer than the adjacent and the opposite sides.

   For an angle $\alpha > 45°$, the opposite side is longer than the adjacent, hence $\tan \alpha > 1$.
2. 63.4° and 27.6°, correct to 1. d.p.
3. 0.63 to 2. d.p.
4. 0.866, 0.577, 1.732, 0.707.
5. Hypotenuse = 4 units, other side = $2\sqrt{3}$ = 3.464 to 3 d.p. This triangle is similar to the right-angled triangle in Fig 3.4, with lengths in the ratio 2 : 1.

Practical Assignment (3)

5.7°

Practice Questions (4.1)

1. 0.243 metres 2. 2.86° 3. 41.4°

Practice Questions (4.2)

1. 3.4 metres 2. 10.3 km 3. 214.5 metres

Practice Questions (4.3)

1. 19.5° 2. (i) 35.3° (ii) 45°

Progress Questions (4)

1. 3.36 metres 2. 87.7 metres 3. 3.5 metres 4. 28.1°

Practice Questions (5.1)

1. 40° 320° 180° 270° 0° (or 360°) 280°
2. π 0 (or 2π) $3\pi/2$ π π $5\pi/4$

Practice Questions (5.2)

1. $\sin 215° = -\sin 35°$, $\cos 215° = -\cos 35°$, $\tan 215° = \tan 35°$
2. $\sin 325° = -\sin 35°$, $\cos 325° = \cos 35°$, $\tan 325° = -\tan 35°$
3.

Quadrant	sin	cos	tan
1	+	+	+
2	+	–	–
3	–	–	+
4	–	+	–

4. + – – – + – + +

Practice Questions (5.3)

1.

	30°	45°	60°
sin	½	$1/\sqrt{2}$	$\sqrt{3}/2$
cos	$\sqrt{3}/2$	$1/\sqrt{2}$	½
tan	$1/\sqrt{3}$	1	$\sqrt{3}$

	120°	135°	150°
sin	$\sqrt{3}/2$	$1/\sqrt{2}$	½
cos	– ½	$-1/\sqrt{2}$	$-\sqrt{3}/2$
tan	$-\sqrt{3}$	–1	$-1/\sqrt{3}$

2. 0.174 –0.985 –0.176 –0.906 –0.966 0.675
   –5.67 0.342 –0.342 0.819 0.268 0.643

Practice Questions (5.4)

2. (i) two (ii) two (iii) two 3. (i) 0, 1, 0, –1, 0
   (ii) 1, 0, –1, 0, 1 (iii) 0, 0, 0
4. a) 23.6°, 156.4° b) 68.8°, 248.8° c) 41.4°, 318.6°

Progress Questions (5)

1. a) 0.342, –0.906, –0.500, –0.342
   b) –0.174, 0.866, –0.940, 0.906
   c) –1.428, 1.000, –1.732, 0.839
2. a) 0.64, 5.64, 6.92
3. $\tan \alpha = \tan (180° + \alpha)$ $\cos \alpha = \cos (360° - \alpha)$

Practice Questions (6.1)

1. $b = 1.47$ cm, 3.94 cm 2. $b = 6.7$ cm, $c = 13.9$ cm
3. $a = 9.3$ cm, $c = 11.3$ cm
4. $\angle C = 69.3°$, $\angle B = 57.7°$, $b = 3.7$ cm <u>or</u> $\angle C = 110.7°$, $\angle B = 16.3°$, $b = 1.23$ cm.
5. $\angle A = 71.4$, $B = 60.6$, $a = 7.4$ cm <u>or</u> $\angle A = 12.6°$, $\angle B = 119.4°$, $a = 1.7$ cm
6. $\angle B = 40.1°$, $\angle C = 29.9°$, $c = 6.63$ cm

Practice Questions (6.2)

1. $\angle A = 54.0°, = 76.0°$. 2. $\angle C = 52.0°$, $\angle B = 50.4°$.
3. $d = 3.8$ cm. 4. 92.1°, 40.9°, 47°. 5. 9.224 km.
6. 115 tiles are needed, so this is above Maria's budget.

Progress Questions (6)

1. (i) 3.8 cm, 7.5 cm (ii) 13.1 cm^2
2. (i) $\angle D = 39.8°$, $\angle E = 111.7°$ (ii) 9.0 cm^2
3. 6.48 cm 4. 9.24 km
5. Yes, the ladder needs to be at least 3.47 m
6. No. Area of one tile = 260 cm^2, she would need 115 tiles

Practical Assignment (6)

The area of the kite is 0.45 m^2. To reduce this significantly would make the kite too narrow.

Practice Questions (7.1)

2. (i) 045° (ii) 225° (iii) 315° (iv) 135°

Progress Questions (7)

1. (i) 320° (ii) 200° (iii) 458 km (iv) 271°
2. (i) 18.2 km (ii) 346°
3. 18 minutes (0.306 hours) 4. 31 minutes (bus 8 minutes + walk 6 minutes = 14 minutes)

Practice Questions (8.1)

1. Degrees: (i) 0°, 360° (ii) 180° (iii) 90°, 270°
   Radians: (i) 0, 2π (ii) π (iii) $\pi/2$, $3\pi/2$

3. Degrees: (i) 0°, 180°, 360° (ii) 90°, 270°
   Radians: (i) $0, \pi, 2\pi$ (ii) $\pi/2, 3\pi/2$
5. (i) 360° (ii) 180°

Practice Questions (8.2)

1. (i) −1, −3 (ii) 3, 1 3. (i) 0, −2 (ii) 2, 0

Practice Questions (8.3)

1. Vertical translation, 3 units up
2. Horizontal translation, 90° to the right
3. Vertical stretch, factor 4
4. Horizontal stretch, factor ½
5. Vertical translation, 1 unit up
6. Horizontal translation, 45° to the left.
7. Horizontal stretch, factor 3
8. Vertical translation, 2 units down
11. (i) Horizontal stretch factor ⅓, vertical translation 2 units up
    (ii) Horizontal translation 30° to the right, vertical stretch, factor ½
    (iii) Horizontal stretch factor ½, horizontal translation 60° to the right

Progress Questions (8)

1. (i) 30°, 150°, −210°, −330° (ii) 210°, 330°, −30°, −150°
2. (i) 45°, −135° (ii) −45°, 135°
3. 107°, 313°
4. 144°, 264°
5. Reflection in y-axis, vertical translation up 1 unit
   (i) 270° (ii) 90° (iii) 30°
6. The graph of $y = \cos(-x)$ is the same as the graph of $y = \cos x$. Therefore, $\cos x = \cos(-x)$ for all x.
7. (i) 3, 1 (ii) 2, 0 (iii) ¼, −¼

Practice Questions (9.1)

(i) $56.3°, 236.3°, 56.3° + (180° \times n), 236.3° + (180° \times n)$
(ii) $12.3°, 107.7°, 132.3°, 227.7°, 252.3°, 347.7°, 12.3° + (120° \times n), 107.7° + (120° \times n)$

(iii) No solutions in the range 0° to 360°, general solution $200.5° + (720° \times n)$, $339.5° + (720° \times n)$

(iv) 67.5°, 157.5°, 247.5°, 337.5°, $67.5° + (90° \times n)$

Practice Questions (9.2)

1. (i) 128.8°, 311.2° (ii) 59.1°, 119.1°, 179.1°, 239.1°, 299.1°, 359.1° (iii) 145.6°
2. (i) 45°, 225° (ii) 30°, 150°, 210°, 330° (iii) 44.5°, 195.5°

Practice Questions (9.3)

(i) 0, 2π General solution $2n\pi$ (ii) 0.45, 2.02, 3.59, 5.16 General solution $0.45 + n\pi/2$

(iii) 0.41, 2.73 General solution $0.41 + 2n\pi$, $2.73 + 2n\pi$

Progress Questions (9)

1. (i) The maximum positive value of $\sin\theta$ is 1
   (ii) We cannot take the square root of -2
   (iii) This equation becomes $\cos\theta = 1.2$, but the maximum positive value of $\cos\theta$ is 1
2. (i) $\cos \frac{1}{2}\theta$ is positive in the range $-\pi$ to π
   (ii) $\sin(-\frac{1}{2}\theta)$ is negative in the range 0 to 2π
3. (i) 22.5°, 112.5°, 202.5°, 292.5°, 67.5°, 107.5°, 197.5°, 287.5° (ii) 300° (iii) 45°, 315°, 135°, 225°
4. (i) $1.11 + n\pi$ (ii) $0.15 + 2n\pi$, $2.991 + 2n\pi$
   (iii) $4.18 + 4n\pi$, $8.38 + 4n\pi$

Practical Assignment (9)

26.8°, 153.2°

Practice Questions (10.1)

1. $(x-1)(x+1) \equiv x^2 - 1$ 2. $x - 1 = 100$ 3. $\tan\theta = 1$
4. $\sin\theta \equiv \sin(180° - \theta)$ 5. $\cos\theta \equiv \cos(-\theta)$ 6. $\sin^2\theta = 1$

Practice Questions (10.2)

4. Addition formula 4

Practice Questions (10.4)

1. (i) 1.11 (ii) −0.38 (iii) 0.69, 2.26, −0.88, −2.45
2. (i) 1.91, −1.95, 0 (ii) 1.57 (iii) −0.52, −2.62
3. (i) $\pi/2$, $-\pi/2$ (ii) $\pi/6$, $5\pi/6$, $-\pi/2$ (iii) $0 + n\pi$, $0.96 + n\pi$

Progress Questions (10)

2. (i) $\sin 3\theta \equiv 3 \sin \theta - 4 \sin^3\theta$ (ii) $\cos 3\theta \equiv 4 \cos^3\theta - 3 \cos \theta$
3. (i) $n\pi$ (ii) $\pi/8 + n\pi$ (iii) $0, n\pi, 0.62 + n\pi, -2.53 + n\pi$
   (iv) $\pi/4 + n\pi$ (v) $\pi/4 + n\pi, -0.98 + n\pi$

Practice Questions (11.1)

1.

x	cos x	sec x
0°	1	1
30°	$\sqrt{3}/2 = 0.866$	$2/\sqrt{3} = 1.155$
45°	$1/\sqrt{2} = 0.707$	$\sqrt{2} = 1.414$
60°	½ = 0.5	2
90°	0	1/0 (undefined)

2.

x	tan x	cot x
	0	1/0 (undefined)
30°	$1/\sqrt{3} = 0.577$	$\sqrt{3} = 1.732$
45°	1	1
60°	$\sqrt{3} = 1.732$	$1/\sqrt{3} = 0.577$
90°	undefined	0

3. (i) 1 (ii) –1 (iii) 1 (iv) –1
4. 45°

Practice Questions (11.2)

1. 45°, 225°, 26.6°, 206.6°
2. 30°, 150°, 270°
3. 30°

Progress Questions (11)

1. (i) (ii) and (iv) can never be true
3. (i) –0.68 (ii) $\pi/4, -3\pi/4$ (iii) $\pi/4, -\pi/4$

Practical Assignment (11)

73.4°

Practice Questions (12.1)

(i) 90°, 323.2° (ii) 31.7°, 260.9° (iii) 57.6°, 159.2°
(iv) 81.4°, 349.2°

Practice Questions (12.2)

1. (i) $\sqrt{20}\sin(\theta - 63.4°)$ (ii) $\sqrt{20}, -\sqrt{20}$ (iii) 153.4°, 333.4°
2. (i) $\sqrt{10}\sin(\theta + 18.4°)$ (ii) $\sqrt{10}, -\sqrt{10}$ (iii) 71.6°, 251.6°
3. (i) $\sqrt{41}\sin(\theta + 51.3°)$ (ii) $\sqrt{41}, -\sqrt{41}$ (iii) 38.7°, 218.7°

Progress Questions (12)

1. (ii) 2, −2 (iii) 60° (iv) 240° (vi) $\sqrt{3}\sin\theta + \cos\theta$ (vii) 0°, 120°

   [These are the solutions which you read from your graph in (v)]
2. 0.26, −2.16 3. 13, −13

Practical Assignment (12)

23.6°, 156.4° (You may not be able to get this degree of accuracy from your graph)

Practice Questions (13.2)

1. (i) $2\sin 35°\cos 5°$ (ii) $2\cos 40°\sin 10°$
   (iii) $2\cos 40°\cos 20°$ (iv) $-2\sin 50°\sin 30°$
2. (i) 1.143 (ii) 0.266 (iii) 1.440 (iv) −0.766
3. (i) $2\sin 3\theta\cos\theta$ (ii) $2\cos 3\theta\cos\theta$
   (iii) $-2\sin 2\theta\sin\theta$ (iv) $2\cos 4\theta\sin\theta$

Practice Questions (13.3)

1. 0°, 60°, 120°, 180°. 2. 0°, 30°, 60°, 90°, 120°, 150°, 180°
3. 0°, 90°, 180° 4. 22.5°, 67.5°, 112.5°, 157.5°, 180°.

Practice Questions (13.4)

1. 45°, 135° 2. 120°, 22.5°, 67.5°, 112.5°, 157.5°
3. 0°, 30°, 150°, 180° 4. 45°, 135°, 90°, 180°

Progress Questions (13)

2. C = 90°, D = 0° 4. $0.1225 \pm n\pi/2$

Chapter 14 Revision Questions

1. a) 66.4° b) 17.5
2. a) 8.60 b) 6.66

3. a) 045° b) 337.5° c) 112.5°

4. a)

Degrees	Radians
180°	π
90°	$\pi/2$
60°	$\pi/3$
45°	$\pi/4$
270°	$3\pi/2$
120°	$2\pi/3$
30°	$\pi/6$
150°	$5\pi/6$

b)

Angle	Quadrant
3.5	3
$2\pi/3$	2
0.96	1
5.9	4
$3\pi/4$	2
1.5	1
$5\pi/8$	2

5. a) 0.47, 2.05, –1.10, –2.67
   b) No solutions in the range
   c) $\pi/3, -\pi/3$

7. (iii) 9.24 m 8GH = 14.4, FH = 16.1

10. a) 82.0°, 231.6° b) 18°, 54°, 90°, 126°, 162°, 234°, 270°, 306°, 342°, 60°, 120°, 240°, 300° c) 45°, 135°

12. a) Inverse cos(0.4) = 6.4° arccos (0.4) = 66.4° $\cos^{-1}(0.4) = 66.4°$ b) (i) 0.36 (ii) $\pi/3$

13. 24.14 km

14. $\sqrt{10}\sin(\theta° - 18.4°)$ or $\sqrt{10}\cos(\theta - 71.6°)$

15. (i) $\dfrac{2\tan 2\theta}{1-\tan^2 2\theta}$ (ii) $\dfrac{3\tan\theta-\tan^3\theta}{1-3\tan^2\theta}$

16. (i) $\sqrt{2}$ (ii) $\sqrt{3}$ (iii) 35.3°

Glossary

Acute angle: An angle in the range 0° to 90°.

Adjacent (side): The side of a right-angled triangle joining the reference angle to the right angle.

Angle: The change, or difference, in direction between two intersecting straight lines.

Arc: A section of the circumference of a circle.

Complementary: Two angles are complementary if they add up to 90°.

Cosecant: The cosecant function is defined as $\operatorname{cosec} x \equiv 1/\sin x$.

Cotangent: The cotangent function is defined as $\cot x \equiv 1/\tan x$.

Cosine ratio: The ratio adjacent : hypotenuse.

Degree: Unit of angle measurement, where one degree is 1/360th of a revolution.

Equilateral: An equilateral triangle has three equal sides and three equal angles.

Equation: A mathematical statement which is true only for particular value(s) of the variable(s).

Harmonic form: An expression such as $R \sin(\theta \pm \alpha)$ or $R \cos(\theta \pm \alpha)$ is said to be in harmonic form.

Hypotenuse: The longest side of a right-angled triangle, i.e. the side facing the right angle.

Identity: A mathematical statement which is true for all values of the variable(s).

Inverse: An inverse function has the opposite (i.e. reverse) action to the original function.

Isosceles: An isosceles triangle has two equal angles and two equal sides.

Quadrant: One quarter section of a Cartesian graph, which is divided into these sections by the: x- and y-axes.

Obtuse angle: An angle in the range 90° to 180°.

Opposite (side): The side of a right-angled triangle opposite to the reference angle.

Radian: Unit of angle measurement defined by an arc equal in length to the radius of the circle.

Reciprocal: The mathematical inversion of a quantity, e.g. the reciprocal of 10 is 1/10.

Reflex angle: An angle in the range 180° to 360°.

Secant: The secant function is defined as $\sec x \equiv 1/\cos x$.

Sine ratio: The ratio opposite : hypotenuse.

Supplementary: Two angles are supplementary if they add up to 180°.

Tangent ratio: The ratio opposite : adjacent.

Index